About the Authors

ılıılıılıılıılıılıılıılıılıılıılıılıılıılıılıı

A five-time winner of the NBA Most Valuable Player Award and a twelve-time All-Star, BILL RUSSELL was the centerpiece of the Boston Celtics Dynasty that won eleven NBA championships during his thirteen-year career. For the final two of those championships, Russell was the first African-American coach of a major-league professional team. He is widely considered the greatest team player in NBA history, father of the modern professional game, and one of the most significant Americans of the twentieth century in sports. He has published three previous books, including the national bestseller *Russell Rules*.

In February 2009, the National Basketball Association honored Russell by officially renaming its NBA Finals MVP Award the "Bill Russell NBA Finals Most Valuable Player Award."

ALAN STEINBERG is the author of numerous books, including the *New York Times* bestseller *Behind the Mask* and *Black Profiles in Courage* with Kareem Abdul-Jabbar.

Red and Me

Also by Bill Russell

NONFICTION

Go Up for Glory

Second Wind

Russell Rules

Red and Me

My Coach,
My Lifelong Friend

Bill Russell

with Alan Steinberg

HARPER

NEW YORK • LONDON • TORONTO • SYDNEY

HARPER

A hardcover edition of this book was published in 2009 by Harper, an imprint of HarperCollins Publishers.

HarperCollins books may be purchased for educational, business, or sales promotional use. For information please write: Special Markets Department, HarperCollins Publishers, 10 East 53rd Street, New York, NY 10022.

FIRST HARPER PAPERBACK PUBLISHED 2010.

Designed by William Ruoto

Library of Congress Cataloging-in-Publication Data is available upon request.

ISBN 978-0-06-179206-9

HB 04.26.2023

Not what we give, but what we share. . . .
for the gift without the giver is bare.
JAMES RUSSELL LOWELL

A friend may well be reckoned a masterpiece of nature.
RALPH WALDO EMERSON

Contents

‖‖‖‖‖‖‖‖‖‖‖‖‖‖‖‖‖‖‖‖‖‖‖‖‖‖‖‖‖‖‖‖

Prologue   xi

Chapter 1:   Common Ground   1

Chapter 2:   From Place A to Place B   19

Chapter 3:   A Pretty Good Sign   35

Chapter 4:   My Father's Son   65

Chapter 5:   What's Best for the Team   89

Chapter 6:   We All Just Lived It   111

Chapter 7:   Boy, What Fun   127

Chapter 8:   Godspeed   143

Chapter 9:   This Is My Friend   163

Epilogue   175

Acknowledgments   185

|||||||||||||||||||||||||||||||||||||||||

"Don't fall."

Red Auerbach and I were friends our whole adult lives—almost fifty years. But we never talked about it. That was part of why the friendship worked. He always knew he was one of the few people I cared a great deal about, and I always knew I was one of the few people he cared a great deal about. It didn't have to be said.

When I joined the Boston Celtics in 1956, I didn't know much about Red Auerbach as a coach, and I didn't care about him. Since my relationships with my previous coaches, in college and on the U.S. Olympic team, were unfriendly, I just expected another adversarial relationship with a coach. And I was completely comfortable with that. But when I arrived in Boston after the Olympics, Red was secure within himself, and I was secure within myself. So he didn't have to prove that he was a great coach, and I didn't have to prove that I was a great player.

Although we came from different tribes as men, we recognized early on that as professionals we had a common agenda: to win basketball games. As our basketball relationship played out, we also realized that, philosophically, we shared the same perspectives on life. Our core philosophies—of how to be men, how to be professionals, how to be friends—were in tune, so we never had to talk about who we were or how to conduct ourselves. We just lived it. Over the next thirteen years, basketball set the stage for our relationship to evolve from caution, to admiration, to trust and respect, to a friendship that lasted a lifetime.

After I retired in 1969, Red and I didn't see each other very often. But, even though I lived in Seattle and he lived in Washington, D.C., we each remained a strong presence in the other's life. A few times a year, we'd end up in Boston at the same time. If the Celtics were home, we'd go see a game together at the Fleet Center. The rest of the time, we kept our relationship current by phone. In 2005, when Red's health deteriorated, I phoned regularly to let him know I was thinking about him. I knew he was getting lots of calls. I also knew that when people aren't doing well, they expend a lot of time and energy telling their friends not to worry. I didn't want to drain Red's energy for my benefit, so I kept the conversations short. All I cared about was that this was my friend. All he cared about was that he was hearing from his friend.

I remember our last conversation, weeks before he died, as vividly as our first championship together half a century ago. He answered the phone with his famous cranky growl, "*What!*"

"Red?"

"Who's *this*?" Still cranky. . .

"This is William F. Russell."

"Hey, Russ! How you doing?"

"I'm all right. The question is, how are *you* doing?"

"Well, you know, it's day to day. Doesn't get any better. But don't worry about it. I'm okay." He didn't want anyone to feel sorry for him—ever. His attitude in life was still exactly like mine: *I can take care of myself. I don't need anybody's pity.*

"Is there any way I can help?"

"Nah. If I need anything from you, I'll call you."

"All right. See you later."

That was it. Short and to the point, like almost all our conversations. Of course, I didn't know it would be our final one. But, though a lot was left unsaid, I have no regrets. We understood each other completely—that was the foundation of our friendship. It's the foundation of all lasting friendships. Years ago, I came up with a saying that expresses it precisely. I don't know whether I heard it somewhere or made it up. But it became part of who I am: *It is far more important to understand than to be understood.*

Before that last call to Red, my wife, Marilyn, and I had already arranged a trip east in mid-October. After a stop in Orange County, California, so I could play in a celebrity golf tournament that raises money for the Susan G. Komen Breast Cancer Foundation, we drove to Hyannisport, Massachusetts, for the Robert F. Kennedy Memorial Golf Tournament. I knew Red's health wouldn't get any better. And my old Celtics teammate Frank Ramsey, who lived in Kentucky, had some health issues from the year before. So I thought it was time to see both of them again.

When we arrived at Red's home in D.C., his housekeeper let us in. Red came out and we hugged, and he led us into his den slowly, using a cane. He sat in his favorite big chair and set the cane on the floor. We started talking about life—where he was in his, and where I was in mine. I said, "You know, I think I finally found a partner." He knew that I meant Marilyn because he'd met her at a Celtics game a few years back. But I hadn't shared much about our relationship. We never discussed our private affairs, unless one of us had a good reason to ask.

He said, "That's great, Russ. I'm glad for you." He gave me a wink. "I know you've been by yourself."

I said, "Well, you know what Dottie used to say about me?" He brightened at the mention of her name. She'd been his wife for fifty-nine years before passing away six years ago.

"Dottie said a lot of things about you. Which one?"

"When I met you both the first time at your reception party, she told you, 'I'm so glad you drafted him. What a nice young man!'"

"Oh yeah," he snorted. "You sure fooled *her*!"

It was Red's usual Brooklyn-Jewish needle—his way of expressing friendliness. We both used to needle each other all the time.

I said, "She was my biggest fan."

"So?"

"So, Red. That's how I feel about *Marilyn*!"

My big laugh rolled into his little one. It was good to see him cheered. Then we discussed the long drive to D.C. He said, "You always did love driving, didn't you? What happened to that sports car you had? You still driving that thing like a maniac?"

He meant my Lamborghini. He used to needle me like hell about the day, when I was coaching the Celtics, I drove it in a snowstorm before a game. It's a very low-slung car with almost no under-clearance, so I ended up in a snowbank. I didn't make the game until the last quarter, and Red had to come out of the stands to coach.

"Nope." I smiled proudly. "We are driving . . . a nice, slow *minivan*."

"It's come to that, huh?"

I knew that it still bothered him not to be driving anymore.

Until he got sick, he always drove himself everywhere in a convertible with the license plate *CELTICS*. I laughed, remembering the time he was scheduled to speak at a school and they sent a limousine for him, but he just followed the limo in his car.

We ended up discussing basketball. He said, "Jesus Christ, these goddamn ballplayers today. They don't know their asses from a hole in the ground. I blame the coaches. They call a time-out and it looks like a mob scene. All these assistant coaches talking to each other. And then, for five or six seconds, to the players. What the hell can you tell a player in five seconds? You ever see anything as dumb as that? You don't need any of these goddamn assistants. You end up coaching *them*, instead of the players!"

He was back in his element. But I didn't want to get him drained, so, finally, I stood up to leave. "Okay Red," I said. "We have to get going." When Marilyn and I reached the door, I told Red, matter-of-factly, "I'll see you later."

As I turned to leave, he called out, "Wait a minute! Wait a minute!" He got up out of his chair and shuffled to us, gingerly, on his cane. "Listen, Russ," he said earnestly, just like in one of our private player–coach meetings. "This is something important. When you get old, don't fall. Because that's the start of the end. So remember: *Don't fall.*"

It caught me off guard. "Okay, Red," I assured him. "I'll do my best not to."

In the past, our saying good-bye was, as we both liked to say, "no big deal." This felt different, and it struck me. For one thing, he'd made a big effort to get up and walk over to me and deliver the message. For another, I knew he'd taken a fall a few months earlier. And when a lot of people get old, they're doing okay and then they fall, and they never recover from it.

When Marilyn and I got outside, I smiled to myself. I realized that when Red said, "Don't fall," it was really a warning for his friend: "When you get old, be careful. Take care of yourself."

That comforted me. He was expressing affection. We never did that openly with each other—it was always unspoken. We both grew up with certain macho assumptions in our era about how men expressed their feelings. One was that a man was always careful not to say "I love you" to another man. Even today, most don't say "I really like you" to their friends, or even "We're friends." So we developed other ways to communicate those feelings—but then we moved on quickly. We never dwelled on that stuff long. We switched to the next agenda.

Once Marilyn and I were in the car, it occurred to me that Red's warning not to fall was something a woman might call "tender"—a tender gesture of friendship. It touched me deeply. As sick as he was, as frail as he felt, he was thinking of his friend. He was thinking of me.

As we drove away, I reveled in that moment, and in our little visit. I thought, "I'm sure glad I did that." But then, pretty fast: "Now, tomorrow morning, let's get back on the road."

Red and Me

Chapter 4

||||||||||||||||||||||||||||||||||||||||||

Common Ground

To me, friendship is simple.

The way I like to picture it is this famous photo of Einstein standing at a blackboard filled with long, complicated equations that add up to a short one at the bottom: $E=mc^2$. The genius was not only that he understood those long equations and what could be done with them, but also his ability to condense them into the short one that defined them. My relationship with Red Auerbach was those complicated equations on Einstein's board. Our $E=mc^2$ was the simple friendship that defined them.

I have a finite number of friends. I keep that number small and it never changes. Someone once asked me, "What happens when you make a new friend?" I laughed, "*Someone* has to go!" The fact is, throughout my life—and it's been a very full one—I've had only a few dear friends. Each one is completely different. But Red was a special one, and probably the most unexpected. Our relationship can't be described by logic. There was no apparent reason we ever should have been friends. We came from different tribes and places: an immigrant-stock, Jewish white guy from New York, and a rural black guy from segregated Louisiana by way of inner-city California projects. Yet we found a way to meet on common ground. That was what made the friendship unique.

I can't speak for Red about what he brought from his tribe, but I can relate some of the key lessons I learned from mine. Starting with the fact that the first thing I knew as a person was that both my parents loved me. Their unconditional love was the most valuable thing I could get from them—it shaped my character and built my foundation. My respect and love for them has lasted my entire life.

Since I came from the Deep South of the 1930s and '40s, it wasn't a foregone conclusion that I would ever be open to being friendly with a white person. But by listening to my family and observing how they conducted their lives, I absorbed core values—respect, integrity, trust, honesty, loyalty, fairness, independence, empathy—that prepared me to be open to Red Auerbach. They were educators that way; they knew what was important in life. And they lived it every day, through example and the spoken word.

My father, whom I always called Mister Charlie, inherited his values from his father, for whom he had enormous respect, admiration, and affection. Grandpa Russell was a tough-minded, hard-working, independent farmer-sharecropper, who kept his own mule team and worked for himself. He was more stubborn than his mules, and he wouldn't tolerate an insult to his dignity or to his family. To me, he was truly a great man—the very essence of manhood. My father told me wonderful stories about his father that I never forgot. These stories taught me, as Mister Charlie always put it, "how to survive in the world, and prosper, with your manhood intact."

One important story was about how, when my father was born, Grandpa Russell realized there were no schools for black children in the area, so he decided to build one himself. He went to the mill, ordered the lumber, and paid in advance. Word of his purpose got out fast. One thing about rural culture back then was that people got together in small groups and talked about everything going on in town. It was like oral text messaging—word shot across the community almost instantly.

When my grandfather drove his mule team and wagon to the mill to get his lumber, the clerk at the mill refused to give it to him. He told Grandpa, "Negro kids don't need no school. They don't need to read to pick no cotton."

Grandpa said, "Okay, Sir. You can give me my money back." My father told me, "Grandpa always said 'Sir,' even when he was fixing to whip a white man's ass!"

The guy wouldn't do that either. He said, "Hell, there ain't no agreement with a Negro that a white man's got to respect."

Grandpa said, "Well, Sir. Then you got three options. You can give me my lumber. You can give me my money. Or I can *kill* you."

Now, in the racially charged atmosphere in Louisiana at the time, that was a way for a black man to get *himself* killed. Grandpa didn't care; he refused to tolerate injustice or disrespect from anyone. Everybody knew about his sense of fairness and his unyielding pride: *I will not allow any man to impose his will on me.* They also knew that he would use his shotgun, if necessary. So Grandpa got his lumber—and he built his school, and raised $42 for the first teacher's yearly salary. Many years later, I went to a school just like it.

Mister Charlie always stressed how proud he was of Grandpa for fighting for respect. Grandpa stood up for himself his whole life. Like the time he scared off the Klan in 1917. The way my father told it, come the end of the crop season a white farmer refused to pay Grandpa his fair share of crop sales in return for Grandpa's help planting, plowing, and sowing the farmer's crops for him. When Grandpa gave him an earful, the farmer threatened to beat him for his insolence. "Sir," Grandpa said. "You and who else?" He got his shotgun and ran the farmer off his land. But not before the farmer warned, "We'll get you for this! Tonight!"

Grandpa went back home—he owned his own house until he died; he was a completely independent man—and moved his family someplace they wouldn't be found. Then he returned to wait with his dog and his shotgun. That night, the dog started barking when a group of Klansmen called Night Riders drove up in trucks. One of them yelled, "Come on out here, old man, and take your whipping, so you can learn how to treat a white man!" Grandpa yelled back, "Sir, you'll have to come on in this house and get me!" When someone fired a stray shot at the house, Grandpa unloaded his shotgun, filled with buckshot, into the darkness. He cut down tree limbs, and nobody ever said what else. But the Night Riders vanished in a hurry; they hadn't bargained for that.

I love that story. As a kid, all I knew about Night Riders was that they patrolled after dark, and they hated black folks and thought they could get away with anything. At that time, in that place, everything was literally black and white to me. I perceived Night Riders as "bad guys" like the ones we saw in cowboy films. Hearing about how Grandpa ran off the "bad guys" thrilled me and made me feel proud, just like my father felt. But the reality wasn't black and white. Most white Southerners never said publicly what they really believed in and were willing to die for. Yet they made it abundantly clear to us black folk, through mass intimidation, what *we* could die for.

Grandpa had a motto, which he told to my father, who told it to me: "A man has to draw a line inside himself that he won't allow any man to cross." I always have been proud of his heroic dignity against forces more powerful than him. And it left a deep impression on me that he would not allow himself to be oppressed or intimidated by anyone. As a young man, I adopted Grandpa's motto and drew my line inside, and I have stood strong behind it my whole life. I have my own motto now: "If you disrespect that line, you disrespect me."

Mister Charlie was a strong, dignified, resourceful man and a devoted husband and father. There's no debating that he was my best friend, period. He didn't tell me "I love you" until he was in his eighties, but it didn't have to be said. I knew it from the moment I was born, because he showed it to me in so many ways. For example, he told me a story about something that happened to him during World War II, when I was nine and he worked at the Bancroft Bag Factory in our hometown of West Monroe.

At the factory, he was basically the superintendent's assistant. He could run all the machines, he was good with his hands, and he was smart enough to figure things out on his own. Anytime the boss had a problem, his response was, "Hey Charlie. Come take care of this for me." But he would not, or could not, put my father in charge. Mister Charlie was one of the few black men working there, and the superintendent told him he couldn't have a "Negro" supervising white men. One day, one of the truck drivers was off someplace and the truck broke down. The superintendent told my father, "Hey, Charlie. Take my car keys and go on down with my car. Give the driver my keys and he can bring it back. You stay there and do what it takes to get the truck running again."

When he accomplished that, he decided to ask for a raise. The superintendent said, "Charlie, I can't give you no raise. If I give you a raise, you'll be making more than some of my white boys. I can't pay a nigger no more than I pay a white boy." He spoke matter-of-factly, as if to say, "No harm meant."

That affected my father profoundly. When he got home, he immediately called a friend who'd left Monroe for a job in Detroit, and asked him, "What's the chance of me getting a job if I come to Detroit?"

His friend said, "Oh, they're just begging for people to work in the Ford plant, Charlie. There's an ad in the paper every day. If you come down here on a Thursday, you'll be working on Friday!"

The next Saturday, payday was noontime as usual—that was when the employees went shopping for next week's groceries. But when Mister Charlie picked up his check, he told his boss, "This was my last day."

"Why? What do you mean?"

"I'm leaving."

"Oh Charlie, are you upset about that raise? Listen, here's what I'll do. From now on, I want you to punch in the clock for two extra hours every week. You don't have to do no extra work. Just punch in two extra hours. That way, I can pay you for overtime and that'll give you the raise."

My father said, "No, I don't think so. This was my last day." And he left. He had already devised his plan: he'd packed his bag at home and bought a ticket on the six o'clock train to Little Rock, the first stop on the long haul north. That Monday morning, by "coincidence," people from the draft board came by our house looking for Charlie Russell. They were going to draft him—unless, they told my mother, he came back to work at the factory. She said he was gone and didn't live there anymore and she didn't have his new address. She saw right through them.

It was 1943. If my father had stayed in town, he would have been drafted into the U.S. Army. Damn the fact that he was ineligible because he was married with kids. But he had anticipated that something underhanded might happen. That was why he left so fast and didn't tell anyone he was going. After that, he told me, "Always have a plan" and "Don't ever drop your guard, because some folks are low-down." He used that exact word—"low-down"—to mean that some white people were capable of doing anything to black folks. Even decades later, Mister Charlie reveled in telling that story. He was proud that he'd outsmarted them and stayed a step ahead, so that by the time they reacted, he was gone. It made me proud too—it exhilarated me. This was my father, whom I wanted to emulate, and he was giving me a sense of dignity and pride in who I am. Not a false pride, like that of a braggart, but a powerful, unwavering, line-drawn-inside-of-you sense of self-respect.

My mother, Katie, also taught me lessons about how to be a man. She always thought of me as the most special human being on the planet. Every day, she told me, "I love you more than anything in the world." Katie always talked to me very softly—she almost couldn't bear to raise her voice. That was because she was extraordinarily protective of me. I was so tall and skinny that people said mean things to provoke me, but she wouldn't stand for that. So, until she died suddenly when I was twelve, I lived in a cocoon. After that, I perceived that there were people in the neighborhood trying to get to me, to put me down, because she wasn't around to protect me. That was how I learned to fend for myself. I developed defense mechanisms against anyone I perceived as a threat, and that made me wary and self-reliant.

One lesson she taught me when I was nine had a tremendous impact. We had just moved across the country to Oakland, California, to join my father in his new life out there. At the time, we were living in the Housing Authority projects. One day, I was outside playing when five kids ran past me, one of whom slapped me on the face going by. It was a message and challenge from the local gang that whatever I'd brought from Louisiana didn't hold much weight with them. When I ran home and told my mother, she grabbed the house keys and tugged me through the projects until we found all five kids. She said, "William, are these the boys?" I practically whispered, "Yes Ma'am." "Good," she said, "because you're going to fight all of them, one at a time."

So I had five fistfights in front of my mother. I was only nine and had never had a fistfight before, so I just swung away wildly, mostly with my eyes closed. I won two, lost three. On the walk back, I started sniffling. Katie looked me in the eyes and said, "Don't you feel bad now, William. You did right. You stood up for yourself like a man. Always stand up for yourself like a man." From then on, I always did—because my mother gave me an enormous boost of self-confidence, dignity, and pride that stayed with me my entire life.

Something else very important about my parents' values stayed with me: their shared sense of commitment, loyalty, and devotion. I didn't know it then, of course, but that would resonate with me a decade later on the Boston Celtics, where commitment, loyalty, and devotion came with the uniform. Back in 1943, when my father went to work in Detroit, he was committed to finding a better life for his family, not just working a better job. The Michigan winter was tough on him. He caught a bad cold that lasted three months, so a doctor told him, "If you stay here, you won't get well. You need a warm climate." So he made some calls and came up with a defense industry job at the Moore Dry Dock shipyard in Oakland, California. He lived in a garage for a time, so it wasn't until he pressed the Housing Authority for suitable housing that we were able to join him there. We had been apart only a few months. I was nine, so I didn't know until later in life that, when he left Louisiana, my father had promised Katie he would send for us. He did just that, even though it would have been easier on him to leave us in Monroe. He never lost sight of his responsibilities as a man, a husband, and a father. He was as devoted to us as Katie was. That lesson wasn't lost on me.

I remember it especially when I think of my mother's illness when I was twelve. I never did find out what she had, but I vividly remember visiting her every day at the hospital, and how she kept telling me and my brother, "Your father has always done the best he can. He's a good man. When I'm gone, it won't be easy for him, so you have to help him. Be good kids. Go to school. Get good grades." I don't think I fully grasped what she meant by "When I'm gone" because, at twelve, you never expect your parents to die. I expected her to come back home.

Here's what I didn't know then: During her illness, she made Mister Charlie promise that, no matter what happened, he would make sure that her boys went to college so we could have "a better life" than they had. She was emphatic about it—she was ambitious for us, especially me. She'd always say, "William, you are going to college, no ifs, ands, or buts!" She considered our education a must; it was not an option. It meant so much to her that, when I was born, she gave me the middle name of Felton, after Felton Clark, the president of a Negro college.

In 1946, when Katie died, my dad set about doing what he'd promised her. He made difficult sacrifices, but he didn't complain. He was completely devoted to her and committed to seeing her dream for us come true, even more so after her death. For example, the year before she died, when the war ended, the Oakland shipyards and steel factories shut down. A lot of laborers were put out of work, including my father. But he didn't let that hold him back. He bought an army surplus Dodge truck, constructed a bed for it, and fashioned five benches for the bed and a canvas hood—like a covered wagon—to protect against the elements. Every morning at daybreak, he drove the truck to the corner of Eighth and Center Street. If black folks wanted work, they gathered there and he charged them a dollar a head to drive them to and from the fruit farms to pick peaches, apricots, apples, cherries, and pears. That was his new business. He'd started it almost overnight, for us more than for himself. That was who he was.

The business grew so fast, he contracted with five other men who also owned trucks and with farmers who needed more workers. Pretty soon, his trucks were ferrying hundreds of laborers to work the farms every day. He also did a little supervising in the fields, so sometimes he made $100 a day. He had fun at it, and he was outdoors, and he was helping other black folks who needed work in tough times. Most of all, he was his own man, living on his own terms—a Russell family tradition.

That all caved in when Katie died a year later. But he shouldered the responsibility immediately by selling his successful trucking business and taking a manpower job at McCullough Foundry in nearby Berkeley, just so he could be home every night for us kids. That was miserable work, especially after the fun and easy freedom of his contracting operation. But the biggest sacrifice was going from earning $100 a day in his trucking business to $14 smelting metals into molds.

Mister Charlie lost serious money, and pretty soon we had almost nothing. But he knew we needed someone at home who loved us, and he figured he'd find some other way to raise the money for us to go to college when that time came.

He never once complained for himself—he kept a sunny view. I asked him once if we were poor. He said, "Nope. We're broke. Poor is a state of mind. Broke is a temporary situation." That was how he approached life: burdens were temporary situations. There was always something better, *if* you had a plan and were prepared.

When I think about my father's burden back then, and the sacrifices he made for his kids, two conversations come to mind that illustrate his character and that still touch me. When my mom died, Mister Charlie was in his thirties, a young, strong, virile man. But if he tried to have a girlfriend, and my brother and I caught sight of her, she was automatically unacceptable. That's just the way young kids are. Years later, when I was an adult, I was visiting with my dad and told him, "I owe you an apology."

He said, "Why?"

I said, "I know that after my mother died, every time you got attracted to a woman, if I liked her, my brother didn't. And if he liked her, I didn't. We might've run off a woman that was really good for you. I apologize for that."

He said, "You didn't do anything. I'd *already* ended up with the best one!" He had remained loyal, devoted, and committed to the person he still loved.

The other conversation took place when I got my first $100,000 contract with the Celtics. My father was getting older but was still working in the foundry. I called him up and said, "Listen, you won't believe this. They're actually paying me a hundred thousand dollars a year to play basketball. The reason I'm telling you about it is because you don't have to work anymore. I know how much it costs you to live, and I can take care of it the rest of the way."

He said, "I don't want your damn money."

"Why not?"

"I got a job. I got my own damn money."

I said, "But that's a terrible job. You work in a foundry. All your pants have holes in them from the sparks. There's dust and metal particles in the air. And you're by furnaces and then you go outside and it's freezing. It's *terrible*."

He said, "I can't quit this job."

"Why not?"

"Listen, son. I've given these people thirty-five of the best years of my life. Now, I'll give them a few of the *bad* ones!"

He would never take anything from me, never let me give him anything. He always said, "I can take care of myself." In his later years, he let me take him to NBA playoff games, but that was it. So, finally, I decided to get him a new car. I went to a dealer and bought a car and told them where to deliver it. That way, I figured, there was nothing my dad could do about it. When they delivered the car, they handed him the keys—"Excuse me, I believe these belong to you"—and left. And there was this brand-new car in the driveway.

He told me later that when he realized what happened, he cried. I had never seen him do that. I said, "I wish I had *known* I could make you cry!"

Still later, he called me up and said, "I'm really pissed off at you."

"Why?"

"Because I just got the first speeding ticket of my life!"

When we all returned to Monroe to bury my mother, Mister Charlie had to make another momentous decision. While we were there, Katie's sisters, my aunts, debated who would take my brother and who would take me. They said, "A man can't raise two children alone." But Mister Charlie said, "No! I promised their mother I would raise them and send them both to college, and that's what I'm going to do. It's not open for debate." When I was old enough to appreciate such loyalty and devotion, I had already absorbed something else that Katie taught me very young: Everything you do in life, and every decision you make, has consequences. And those consequences, good or bad, are your responsibility, and yours alone. "That's what it means," she used to say, "to be a man."

All these lessons I learned as a kid—respect for others, self-respect, self-reliance, honor, pride, dignity, commitment, loyalty, devotion, responsibility—were family traditions. So, as an adult, all I did was carry forward those traditions. My upbringing was a key reason why Red Auerbach and I were able to become friends. We brought the same things to the relationship. I am convinced that if we had brought any less, or any more, it wouldn't have worked.

Chapter 2

From Place A to Place B

I was born into the segregated South. My father taught me that, as a matter of survival, black men had to understand white men, but white men did not have to understand black men. That's the way it is, if you're a minority in your own culture. So when I ventured into the world beyond my community, along with my core values I carried certain racial lessons from my tribe. Some people call it instinct. It's not instinct—it's learned.

I never told Red any of my stories of encountering racial ignorance. "My sob story is worse than your sob story" was not part of our conversation. But to understand how any conversation between us was possible, it's important to know about my frame of reference coming out of college. My experiences there set the tone for what I thought I had to look forward to as a professional.

When I attended the University of San Francisco, there was always something going on between the races that, let's say, caught my attention. I knew, from my family, that I did not have to acquiesce to it or accept it, just understand it. All these incidents were about people's preconceived notions of each other. In one way or another, it happens to everyone. You're just going through life when, suddenly, people start imposing their agendas on you, with no regard to you. My father had a catchphrase for this: "little red wagons." He told me, "When people have their own agendas, your attitude has to be: 'That's *their* little red wagons. That's *their* demons they're dealing with. They're just trying to dump them off on me.'"

People dumping their little red wagons on me was one of the two major conflicts I had at USF. I thought I was just another student, but I ended up having more than my share of confrontations. The college authorities consistently prejudged me as the guilty party, a troublemaker, someone bucking the system—always without a hearing. But what troubled me most was that they kept commanding *me* to apologize, instead of the actual perpetrators. Now, why was that?

Back then, USF was a small, all-male Jesuit school with only nine black students on campus. My extended time there was the first I had ever spent in a virtually all-white atmosphere. There were a lot of students from small towns dotting the San Joaquin Valley who brought their religions and cultures and prejudices with them. Many of them had never interacted with black folks before, and some were afraid of us for whatever reasons. Sometimes, that fear erupted into confrontations. Similar encounters had accumulated since my childhood, and by the time I arrived at college they had taken a toll on my patience. To be blunt, at twenty-one I had become a psychologically aggressive person who wouldn't take crap from anyone. *Keep out. Danger zone.*

A typical example happened in the dormitory my junior year. A hostile character approached me and said, "Hey boy." He'd crossed the line already. "What do they call you?"

"My given name is William Russell." My mother used to tell me, "Now William, don't you let anybody call you out of your name." I remembered that—and how she taught me to stand up for myself like a man.

"No," this kid said. "I'm not going to call you that."

"I think you should."

"I think I'll make up a nickname for you."

I warned him, politely but very firmly, "Don't *do* that."

He walked away. A little while later, he came back and baited me. "I got it!" he said. "I'm calling you *Snowball!*"

He woke up on the floor. I mean, I clocked that kid really hard. He tried to clear his head. "Why'd you hit me?"

"Well, I told you not to do that."

I wasn't particularly angry—this kind of disrespect was all too familiar to me. But the fact that he had to ask why he got hit said something about his frame of reference. To understand it, you had to understand the psyche of the nation. At that time, white folks generally felt it was proper to say anything they wanted to black folks. Where I grew up, this behavior was unremarkably consistent. So you had to find a way to handle it like a man. I often thought of something Mahatma Gandhi said: "I do not concern myself with being consistent in what I have said, but with being consistent with the truth as it reveals itself to me." That always resonated with me. No one is completely consistent—we're all human beings with the same basic flaws. But the world changes, and you can't stay the same when things change around you. At some point, you have to engage yourself in the process, learn what it's about, and evolve with the changes. You expect others you respect to do the same.

When I was fourteen and living in the Oakland projects, there was a baseball incident that caught my attention. It involved Brooklyn Dodgers shortstop Pee Wee Reese, a white man from Kentucky, and Dodgers second baseman Jackie Robinson, who had broken the color barrier in the major leagues. In one game, the fans and opposing players screamed racial slurs at Robinson. Finally, Reese strolled across the field, draped his glove over Robinson's shoulders, and chatted with him as if they were a couple of buddies hanging out on the corner. Reese's gesture became legendary, because here was this Southern white man making a bold, public show of support for a black man who so many people hated and feared, just because he was black.

I think that what Pee Wee did was admirable. But, in my era, I never would have allowed anyone to put his arm around me like that, as if to say, "You're my teammate and we're okay." I would have rejected it because I did not ever need, or want, a sponsor or protector. Jackie always said it was one of the most important gestures in his career. That's okay—it was rare to him, it was new. But when I reached the public stage, that ground had already shifted under everyone. Jackie had taken care of a lot of that stuff before I arrived. He moved us from Place A to Place B. So you could no longer treat us as if we were still in Place A.

This was a key part of *my* frame of reference before I got to college. But while Jackie was still provoking change in baseball, that train was overdue at USF. As far as I could tell, the general attitude there was that if you were black, no matter what someone did to you, it was up to you to walk away. Well, who decided that? I was always aware of what I'd be facing, but there were things I wouldn't let anyone do to me.

My roommate K.C. Jones had a very different attitude. When rude or insensitive people confronted him, K.C. just said, "That's ignorant," and deliberately walked away. He was a very good man who gracefully avoided confrontation, although he could have easily beaten the hell out of any of his antagonists if he had wanted to. I must have heard a hundred times during my freshmen year, "Why can't you be more like K.C.!"

My other major conflict at USF was in basketball, specifically with my coach. I want to be clear that he was a good,
decent, fair-minded man—in fact, we later became good
friends. But that was not the issue at the time. The issue
was that he simply could not coach me. He didn't respect my
knowledge of the game and he didn't, or couldn't, perceive
the level of my skills.

Our main point of conflict was that he wanted me to play
exactly like his center had played the previous year. In that
era, virtually every coach in the country subscribed to the
conventional prototype of how a center should look and
play: lumber to a spot under the basket and plant himself
for layups and rebounds—basically an offensive-oriented
game. The accepted wisdom of the day on general defensive strategy was this frozen idea: *A good defensive player never
leaves his feet.* Well, that mindset eliminated me off the bat.
I had invented most of my game on the playgrounds, and
I approached it from a defensive premise that was largely
vertical. I left my feet a lot.

That kid who had preceded me at center was a very nice
guy, but our assets couldn't have been more different. I was
six foot ten, two hundred pounds, with quickness, agility,
and impeccable timing—all of it self-trained. I was also on
the track team. In basketball scrimmages, I could literally
run backward faster than he could run forward. As a very
competitive high jumper, I could touch the top of the back-

board, if I needed to, and be looking *down* at the rim. He was about six foot six, two hundred fifty pounds—with a run and a jump and tremendous effort, he could sometimes touch the net. So my thoughts—which I kept to myself—were, "There are only six teams in our league. In this kid's last year as a starting center, he did not make First Team All League Center or Second Team or Third Team—he made Honorable Mention. His teams never had a winning record. So, why would you make him the role model for how to play the position when he's gone and I'm here, and my game is infinitely more effective?"

Coach was determined to make me play the way he wanted me to, and I was just as determined to play the way I wanted to. When a play developed, I played horizontal and kept my feet on the floor. But when it reached a point where I knew I had to adjust, I would go vertical to block a shot and shut off the other team. I tried to disguise those moves so I wasn't always pissing on the coach's leg. One of my innovations was the blocked shot, which revolutionized how defense could be played. People still ask me today, "Who taught you how to do that?" Well, nobody! I had never seen a blocked shot in a basketball game before I did it! I estimated that, in college, I *averaged* at least fifteen blocked shots a game. Often, I started fast breaks off those blocks, which revolutionized the way offense could be played.

Coach didn't see any of that. My first varsity game, I blocked my opponent's first six shots. Coach pulled me aside and said, "You can't play defense that way!" and he showed me, again, how he wanted it played. I followed his method and my opponent proceeded to score three consecutive layups on me. But coach just told me, "That's the right way to play," and he insisted that I play that way for three years. So what I had to live with was that he never appreciated my contributions. For example, when we were ranked Number 2 in the nation and Kentucky was Number 1, we had a game against Santa Clara in San Jose. If we won that game, we'd be Number 1. Just before halftime, they were ahead by 15, but I put up a shot from about half-court and it went in. That gave us a boost. At halftime, I didn't hear anything that was said in the locker room. I just remember how badly I wanted to help my team win and be Number 1, and telling myself, "Okay. I'll play as hard as I can possibly play and make sure we win this game."

Second half, I scored the first 14 points. They were all hustle points—layups, dunks, tip-ins—and we caught up and won. I outscored their whole team the entire second half. After the game, they let the press into our locker room and I overheard my coach tell a reporter, "It was a complete team victory. No one player was outstanding." That was when he lost me for good. I remember thinking, "He's caught lightning in a bottle and doesn't know what it is or what to do with it." I didn't really blame him. He was a good coach who had some success with his approach, but he couldn't shake his preconceived notions of how the game should be played. That was typical then. Plus, no one had ever seen my unorthodox style before, so no one knew what it was. I had to dismiss his method from my mind and find ways to develop myself on my own. I wasn't trying to create conflict—he knew I respected him as a man. I wasn't trying to draw attention to myself, either. I was trying to improve my team's chances of winning basketball games. *Results* were all I cared about.

I think it worked. My junior year we were 28–1 and the best defensive team in the nation. On offense, we averaged 72 points per game while our opponents averaged 52. I averaged over 20 points and 20 rebounds a game, and I was the only guy in college blocking shots. At the end of my junior season, in 1955, we won our first national championship and I was voted Most Valuable Player of the Final Four. It was funny, though, because a lot of people still didn't think I was very good. When we returned home from the finals that year, sportswriters named another center, Kenny Sears, as Player of the Year. He wasn't bad—he went on to play a little for the New York Knicks. But I was enraged.

League officials knew how shaky their selection was, so they told my coach to have me give a congratulatory speech for Kenny Sears at the awards banquet. Otherwise, how would it look with the year I had, leading us to the NCAA title while Kenny Sears's team finished third in our conference? They even wrote it out for me to say, word for word: "Congratulations, Kenny. You deserve this award. I hope I can get it next year." I refused—it was dishonest and disrespectful to me. From that moment on I decided, "I will never again care about other people's critiques of my performance. All I will care about is winning games. If I win every game I play, that's a fact and nobody's opinion. When they look back, all they'll be able to say is, 'Well, he didn't play the way he should have. But he sure won a lot.'" My senior year, we went undefeated, making it 55 in a row over my last two seasons, and we won our second consecutive national championship.

My difficulties with coaches continued through the 1956 Summer Olympics in Melbourne, Australia. There were no major confrontations; let's just say the Olympic basketball coach did not treat me with respect. In fact, I can't recall ever having had a conversation with him the whole time we were together. He shared the same preconceived notions as almost every other coach about how a center should play. Through that clouded prism, he decided that I was only the seventh- or eighth-best player on the team, and easily the second-best center. So I guess there was nothing to talk about. His one-size-fits-all coaching method was that he was the coach and he knew everything, "so you guys will do things the way they're supposed to be done." Fifty years ago, those kinds of coaches called themselves "traditionalists." They're still around, only today they call themselves "purists." I just called them *mistaken*.

I raise these particular events not as illustrations of what happened to Bill Russell, college student. I raise them to explain the low expectations that I brought to my first interactions with another coach in another world at Boston.

Red Auerbach started coaching the Boston Celtics in 1950, when he was thirty-two and I was a sixteen-year-old junior at all-black McClymonds High School in the West Oakland projects. We had no idea each other existed; we couldn't have been farther apart in every way.

The first time Red heard the name Bill Russell was in 1955, my junior year at USF, before I led our team to the first of our two national championships. In December that year, he got a call from Bill Reinhart, his old college coach at George Washington University. Red had learned a lot about coaching from Reinhart, including how to run a fast break, which wasn't yet in vogue in the pros, and had enormous respect for his judgment. Reinhart had just seen me play in a holiday tournament in Oklahoma City, in the middle of our two-year winning streak. He reported to Red, "This is the guy you want. This kid will make you champions." He didn't explain much—he was a man of few words. But, like Red and me, he knew what was important.

Red didn't take just Reinhart's word. From the day he'd started coaching, all he thought about was winning games. Like me, he was strategy-minded, innovative, result-oriented. A big part of his genius was that he was constantly calculating new ways to improve his team and beat yours. Like my father, he always had a plan. In the mid-1950s, Red's main plan was to draft a big center. The most dominant pro team of the late 1940s and early 1950s was the Minneapolis Lakers. They featured a six-foot-ten, highly skilled center named George Mikan, who was overwhelming—he led them to five titles. Winning a championship—in other words, being the best—was what compelled Red: *Mikan's a winner. I need a winner to win against Mikan*. It made sense.

When I was draft-eligible for 1956, Red still didn't know much about me. Since Reinhart had recommended me so highly, Red tracked my senior season, so he knew I led USF to another championship. But he never personally saw me play. There were no game films back then, and you did your own scouting any way you could, mainly through word-of-mouth. So Red asked his coaching associates and some former players who'd seen me play what they knew about Bill Russell.

One of the people Red spoke with was a former Celtics player named Don Barksdale, who was living in Oakland, California. Before playing pro, Barksdale had attended UCLA, so he'd seen me play. His report duplicated Reinhart's: "This is the guy you need." Ultimately, Red decided I was the type of player who could help him win a championship. He started devising an ingenious plan to maneuver two other teams out of their draft picks so he could grab me for the Celtics in the draft.

It wasn't easy. His first problem was the territorial draft. That was an automatic pick from your team's geographical region. The NBA guys in New York put that rule in because they believed that all the great players were in New York. Red took Tommy Heinsohn from Holy Cross as his territorial pick, but he wasn't slated to pick again until Number 7. The Rochester Royals had the first pick, so the next thing Red did was invent a way to prevent Rochester from taking me.

There were other pressing concerns. One was a private promise he'd made to "Easy" Ed Macauley, his All-Star center. Macauley hailed from St. Louis, where he had a sickly son at home. He was seriously considering retirement to spend more time there with his son. Red promised that he would take care of that problem, although he didn't say how. Since the St. Louis Hawks had the Number 2 pick, Red cooked up something to address Macauley's personal crisis as well as his own ultimate goal of landing me. What he did impressed me when I learned about it a couple of years later.

First, he called Hawks owner Ben Kerner, a man he'd once coached for and despised. If he could maneuver Kerner off me, that would suit Red fine. So he offered up Macauley, his best shooter, and the Celtics' seventh spot in the draft in return for St. Louis's right to pick second. But Ben Kerner knew he had Red over a barrel. So he demanded more. He wanted to throw in Cliff Hagan, a promising young forward from Kentucky. At that juncture, Celtics owner Walter Brown told Red, "You can't trade Ed Macauley! He's our best player!" But then Red had Ed Macauley tell Walter personally, "I *want* you to trade me to St. Louis. You'd be doing me a favor, because then I can take care of my son and still play pro ball." That was enough for Walter; he wore his empathy on his sleeve anyway.

St. Louis was satisfied. Now for the coup de grace: the Rochester Royals. They had first pick, so what could Red possibly do to keep them from saying, "Rochester selects Bill Russell"? The Royals already had Maurice Stokes, a great young center who was leading the league in rebounds, so Red figured they didn't need another big center like me. So he persuaded Walter Brown to call Rochester owner Les Harrison with an unusual proposition. Walter said, "Listen, Les. I'm the president of the Ice Capades. If you lay off Russell at Number one, just pick the date and I'll throw the Ice Capades in your building for two weeks." This was Red's almost compulsively innovative genius working overtime. He knew that in the off-season back then, a lot of those big arenas sat empty. Having the Ice Capades in your building for two weeks was like having the Harlem Globetrotters:

guaranteed sellouts every night—and you've made a profit for the year! Of course, Harrison bit. He was a businessman, and for him, this was good business.

Why do I tell this story? Because, even though I only heard about it a year later, when Red and I were still getting a feel for each other, it gave me key insights into his character that found their way into our relationship. Like me, Red had an unwavering commitment to win at all costs, and he was willing to buck current conventional wisdom to suit his own vision of how to win. Also, he trusted and respected the advice of his friends and had enormous confidence in his own instincts. Finally, he used innovative, result-oriented maneuvering to see his plan through. The fact that he weaved all these different threads together successfully was admirable. At that point in our relationship, it added more mortar to seal our connection.

I met Red for the first time right after the draft, when the Olympic basketball team played an exhibition game at the University of Maryland. He invited the team over to his house in D.C. for a postgame reception. I was a little apprehensive because I played crappy that night—which was rare—and I knew that Walter and Red were watching me. I introduced myself to both of them, saying, "I played really shitty tonight. That is not the way I normally play. I am much better than that. It will *never* happen again." Years later, Red informed me that he'd told Walter that night, "I never heard anything like that from a player before. They always make excuses like, 'It was the coach's fault' or 'They wouldn't give me the ball.' Maybe there *is* something here." He said they were both more impressed by that than the way I'd played.

I didn't have an opinion about Red that night. He was cordial and receptive to me. But my defenses against coaches were already formed. When I had left college, if someone I respected talked to me, I was sensitive enough to listen, digest the information, and analyze it. But anything that other people told me, I dismissed out of hand. At that moment in time, I thought of myself as an intellectual pachyderm—very thick-skinned. I felt I could handle anything anyone threw at me. I knew who and what I was as a man, and I remembered my lessons about how to survive, and prosper, in a hostile environment with my manhood intact. So, by the time I arrived in Boston, my attitude was: "I am a man. I can take care of myself."

Chapter 3

A Pretty Good Sign

Then and now, friendship has always been the most important thing to me. I have encountered a number of people I might have liked to be friends with who fell short of my very tough standards. They just didn't know how. I sometimes kid with people, "In order to penetrate my little circle of friends, you would need the perfect storm of my requirements!" Against the odds, Red Auerbach was one of the few people with whom I developed a friendship that was special. And the key to our connection was that, before we met, we both knew how to be a friend.

Some people think that friends have to be equals in everything—in other words, what I get from you must be equal to what you get from me. That is not necessarily so. What you need that I can provide, and what I need that you can provide, may be very different. Friendship depends on what you give to each other, not what you get from each other. But there's no magic formula. At best there are some guideposts to steer by.

One of the most important is learning how to let someone be your friend. Sounds rather simple, doesn't it? But it's not that common. It takes the desire and commitment to learn all you can about someone, an appreciation for your common humanity, and the ability to allow that person to be who he or she is, without conditions.

I was talking to a close friend who was having problems in his marriage. Later, I happened to talk with his wife, and she mentioned, "We're having difficulties." Since I already knew about that, I said, "Can I ask you something personal?"

She said, "Sure. Ask away."

"Does he love you?"

"Yes, he does."

"You have any doubt about that?"

"Nope."

"How about the other way around—do you love him?"

"Yes, I do."

"Any doubt about that?"

"Nope."

I said, "Well, there's one thing that you have to do. And only you can do it."

"What's that?"

"You have to permit him to love you."

She said, "I'm not sure what you mean."

"He can't love you unless you let him." She still looked puzzled. "He can't love you *unless you let him*! Now, if you think that for him to love you, it means he has to change who he is, that won't work. You have to let that guy that loves you love you as he is. You can't make him over."

She said, "I've never heard anything like that."

While she thought it over, I kidded her a little: "It's like this saying a friend of mine has about marriage: 'Women marry men hoping to change them. Men marry women hoping they will never change.'" We both laughed. I know this little saying about marriage is a male cliché, and I'm sure it applies both ways. That's why I like to greet my married acquaintances, male and female, by asking, "Is the make-over finished yet?" If you can allow yourself to accept and enjoy someone for who they are, without trying to control or change them, you can establish a lasting relationship.

Very early on, I recognized that Red appreciated me this way. He was older than me and he was my coach. But one of the first things about him that I appreciated was that he never implied that he felt he had to take care of me. That would have offended me deeply. Contrary to what many people still believe, he did not project himself to me as a mentor—there was absolutely none of that in our relationship. He also never hinted, or said anything, about being a father figure, such as "You're like a son. I'm taking care of you." I never thought of him that way either. That would have been disrespectful to my father.

I was a man when I arrived in Boston; I wasn't a kid who didn't know anything. And I had the best father a guy could ever hope for. I didn't need a father to take care of me, or a big brother, a mentor, or a sponsor. I wasn't even looking for a friend—our friendship just happened. I needed a coach. Red recognized that right away. Most important, I perceived that he accepted me for who I was, without needing to bend me to his will, or change me, in order to be my coach. That was the fertile soil in which our friendship would eventually thrive.

The language of our friendship was muted. I rarely use the word "love" because that word has been diluted in our culture. I say "affection" instead. Red and I communicated our affection for each other through what we did, not what we said. For thirteen years, the platform for this was basketball, which set the stage for our friendship.

I returned from the Olympic Games in November 1956. By the time I signed with the Celtics, I'd already missed training camp and the first twenty-five regular-season games. I had had to sit around for three weeks without touching a basketball, which concerned me. When I finally arrived in Boston, I was anticipating another knotty relationship with a coach. My attitude was still guarded and aggressive: "If you see me in a fight with a bear, pour honey on me because the *bear's* the one in trouble!"

I wondered about playing with teammates who didn't know me, and when I would play my first game, and how that might turn out. I watched one game in Boston—the first time I ever saw the Celtics play—and then we went to New York, and on a Friday night there, I watched another game. Then Red told me that I'd be making my debut in a televised game that Sunday. After so much inactivity, I didn't think I would look very good because I was out of shape.

When I suited up on Sunday, I was a little nervous—I had a lot on my mind, which, of course, I kept to myself. But maybe it showed, at least to Red. When we started down the corridor to the arena, where I was about to be introduced, for the first time in uniform, to my new teammates, Red said to me, "I heard you can't shoot. You worried about shooting?"

I said, "Not much. But it's been on my mind."

He said, "Well, tell you what. Let's make a deal today, right now. When we talk contract down the line, I will never discuss statistics. All I'll discuss is if we won and how you played. That's all I care about. Don't worry about being a big scorer— I don't give a damn about that. All I want you to do is what you've always done. Play your game. And I won't tell you how to do that. Just play the way you know how."

It surprised me. The past three and a half years, no other coach had shown me such wisdom or consideration. It also left an impression because coaches back then never talked so candidly to their players. This demonstrated a giant leap of faith in me. I mean, here's this guy from the other side of the planet who joins your team, and everyone's wondering, "Can he play?"—because by all the conventional standards, he can't do anything! I was a mystery to everyone—*nobody* knew what to make of me. Red hadn't ever seen me play in college. Everything he knew about me was secondhand, and the reports came back: "*He can't shoot.*"

But that was the beauty of the way Red conducted himself. The past was of no concern. It turned out that Red knew, like most bright people, that you should always find out how a person came by his reputation—who gave it to him and why. Especially in sports, because a guy could acquire a false reputation and it might stick for a whole career. Red was not interested in reputations. His only concern was: "What can this guy do for our team today?"

Of course, I didn't know this about him yet. So, at that point, I thought he expected, basically, someone who wouldn't need many shots, maybe five or six a game, but who was a decent rebounder. The NBA hadn't added "blocked shots" to the equation yet—there wasn't even a stat for that. Nobody knew its significance yet. Probably not even Auerbach. Yet he'd trusted Bill Reinhart's judgment, and so he had put his trust in me.

Red's little hallway bargain with me communicated a few things very clearly. Number 1: He was doing away with the accepted, preconceived notions of what a center, in general, was expected to do. Number 2: Instead of trying to bend me to do what he expected me to do, like other coaches, he was allowing me to do what I expected *myself* to do. In other words, "Be yourself. I accept you for who you are." Number 3: He not only trusted that I knew what I was doing, but he freed me to do the one thing nobody else had let me do since high school: play my own game. This was our first exchange about my game, and he was already revealing his attitude to me: "I don't know enough about you. But I will take the time to learn and we'll do this together." I was still reserved—getting used to this would take time—but I appreciated his willingness to be so open. I thought, "Refreshing contrast."

There was another big contrast in my very first game. Early in the first quarter, the referee called goaltending on me. It was not goaltending—I can see it to this day. I found out later that Red hated bullies, just like I do—bullies are people I have absolutely no use for. Well, Red perceived that ref as a bully. So he came out screaming and cursing this guy, as he would say, "to beat the band." He drew a technical foul. All this on behalf of a rookie. So, after the game, I made it a point to thank Red in the locker room because that was the first time a coach had ever stood up for me like that. I said, "I want to thank you for backing me up."

He spit out a piece of soggy cigar. "Don't thank me," he growled. "That's my job. I can't expect my players to fight for me if I don't fight for them. Besides that, if every time they make a tough call against us I raise holy hell, they might think twice and change their call because they'll know I'll be right back up their ass!" I learned right then that, just like my grandfather and father, Red always had a reason for everything he did. I didn't know at the time that this would come into play later, in a couple of unfortunate incidents down South. But his standing up to the ref in that first game told me I might be able to trust him.

That was a promising first step in our relationship. But I was still skeptical. And I was still conditioned by my own preconceived notions about what to expect from a coach. I shouldn't have been—preconceptions were the very thing I'd been fighting against for years. But there was just too much recent negative history to overcome in one moment. So when things didn't turn adversarial immediately, and Red seemed prepared to meet me on my own terms, it wasn't so much that it was a welcome surprise; it was more like I was willing to give it some time. I thought, "Okay, this is a pretty good sign. But now let's wait and see if the other shoe drops."

Red and I were two very different people. Yet we met on common ground, in a field where we were both highly competent. We did not have to outdo each other. We understood that we had to work together because our success depended on one another. Red had to do things his way, and I had to do my things my way. Fortunately, it was never "either-or." In fact, if he had not been open to the way I played, and the way I conducted myself as a man, I could have ended up playing for several different teams. But we both understood that if I succeeded, it enhanced him, and if he succeeded, it enhanced me. Ultimately, our real success was our ability—without ever having to express it openly—to collaborate in every way we could think of, to help the *team* succeed.

Within a week of my arrival, Red perceived that what I was doing on the court was about more than just proving I could play. He saw that despite all the accepted wisdom about how a center should play, I did not play that way. I was going to be the dominant defensive force in professional basketball, and we both knew that. Red had been an offensive-minded coach. Most similar coaches would have tried to shape me to their offensive scheme and force me to play the way previous centers played in that scheme.

Red didn't do that. Instead of clinging to a fixed strategy, as he learned more about my game he made adjustments in his approach to accommodate my strengths, in particular defensively. It was a compliment to me, but it was the team that reaped the benefits. It intrigued me also that he wasn't even concerned about the many unusual nuances of my game that I'd spent years developing, skills that had threatened or baffled my last two coaches. He focused on results, and, watching me play, he concluded that my results were no accident.

Red was consistent with this adjust-as-we-go approach. Six years before I got there, when Red picked up Bob Cousy in a special draft, they had an interesting—and for me, revealing—exchange. Coming out of Holy Cross, Cousy was a popular local hero known throughout New England for his spectacular ball-handling skills, including behind-the-back and through-the-legs passes that no one else could do—except Marcus Haynes on the Globetrotters, and that was strictly entertainment. During Cousy's first year on the Celtics, he made a lot of those dazzling circus passes. At one point, he decided to ask Red if he was okay with them. Red said, "Cooz, I don't care how you pass the ball. You can pass it through your ass if you want to. Just be sure somebody catches it." He saw only results—he didn't care how you got them.

To say that Red Auerbach was supremely confident would be an understatement. He loved to play gin rummy. I can't think of anyone who won a lot of money off Auerbach in gin. The reason was that he was a mathematician of the highest order. He understood the efficacy of equations, especially in terms of problem-solving. Instead of numbers, in basketball Red applied human elements: *Take these human elements, and*

put this together with this, and that together with that, and here's
how we get another win.

An example: As the season wore on, Red started resting
me for exactly two minutes at the end of the first quarter. I'd
go sit at the end of the bench, but, invariably, he'd yell to
me, "Russ! Get your ass over here next to me!" He wanted
to share what he saw on the floor, from his perspective on
the bench. He never explained his motive, but I knew what
he was doing. He was giving me input I couldn't get from
playing center—in other words, helping me to improve my
game without trying to command or coerce me to fit into a
rigid system.

It was no ordinary few minutes when you sat next to
Auerbach. It was like being at a 3-D movie: everything came
at you so fast, you could barely react. One night, as I was sitting next to him, one of our guys scored a basket. When the
other team in-bounded the ball, Red said, "Shit!" That got
my attention. I said, "What? Nothing's happening." He said,
"That guy's got the shot!" I looked up to see one of our opponents running down the opposite sideline, and sure enough,
they passed him the ball and he got the open shot. Red saw
it coming even before it hatched. I said, "Now, what the hell
did you see there?" He said, basically, "This is what this
guy does, this is what that guy does, this is what the defense
does, so that guy will be wide open." Throughout our games,
he was constantly calculating these things in his head—not

for fun, but to help us find more ways to win. Nothing could have suited me more.

On a similar occasion, I was sitting there and Red grunted, "Shit! We'll lose this one one-twenty-five to one-twenty-three!" Well, the game had just started—we were maybe four minutes in. When it was over, the final score was 125–122. I said, "Goddamn it, Red. How the *hell* did you know that?" He said, "In the second half, when fatigue sets in for both teams, here's what happens. When guys get tired, the first thing they do is start taking it easy on the defensive end . . ." and he went on and on. He had calculated the effects of fatigue into this complicated equation for the final score, and, even with forty-some minutes to go, he came up with an *Auerbach = mc²*.

I was developing an admiration for not only how Red calculated game situations but also how he treated his players. His concern wasn't about controlling us. It was about helping us to improve as teammates. He set that tone by always treating everyone on the club as fellow employees, each due the same measure of respect. One way he manifested this respect was by opening his pregame talks by asking us what *we* wanted to do on offense that game. This was something else new to me, and another part of Red's genius that encouraged my trust. It was funny, though, because Red's public image was that he was so domineering, he made Vince Lombardi seem like a choirboy. But he wasn't like that. He was there to coach men, and he knew the functions and responsibilities of coaching men. He was tough as hell on us when he felt we needed it. When a reporter once asked him, "How do you get along with all those stars?" he said, "There's ten of them and one of me. Let *them* get along with *me*!" That would've been my attitude, too. But even when he was being tough on us, he always asked for our input. In fact, what most people never knew about Red was that he respected that everyone on the team knew as much as he did.

When he talked to us about game strategies, he used to tell us in the huddle, "I don't know everything. I can't put this stuff on the floor for you. You guys have to do that and help me out. But this is what I think we should do. What do you think? It's important to know what you guys think because it's no good if it's a strategy you don't buy into." To me, that was a novel approach. First of all, a coach doesn't know something? Back then, coaches never admitted that—it was always the player who didn't know something. Second, asking us what *we* think? Unheard of. This was not Red being just an interesting coach; it was him being an interesting man. It was perfect for me because I always gave my honest opinions anyway, which had never gone over great with my other coaches. But now they were being solicited!

This was another way that Red learned about his players and what each could do for the team. He watched me very closely, of course, but he wasn't sure what to make of me early on. I didn't speak to him about this. I did not explain my method to him—I just went out and executed it. One of the keys to this relationship was that we learned a lot about each other through observation, talking, listening, and osmosis. There was probably more integrity and substance in that method than you can get through conversations.

After a while, observing Red, I knew a couple of important facts. Red's agenda was that he would do anything he could to win games for his team, and so was mine. I realized that I could help him fulfill his agenda, and he could help me fulfill mine, if we worked together more closely. Red respected every player as he was, and listened to each individually. But he did not, strictly speaking, treat everyone equally. He recognized that everyone was different—we all had different contracts, different physical and mental abilities, different mindsets and habits and methods of preparation. He also had to consider all of our different agendas. With professional basketball players, there are never enough minutes and never enough shots to satisfy everyone in every game. We didn't start out as equal ballplayers, so that kind of "equal" is not helpful in a team circumstance. Red knew this.

He was a brilliant mathematician and a brilliant psychologist. As a psychologist, Red would talk to a guy by mainly listening. That was how he got more information about his players and their agendas. I saw him do that so often that when he was talking to me, I always knew what he was doing. He wasn't assessing my needs—they didn't come into play. He wasn't asking himself, "What's this person all about?" He was asking, "How can I help this guy contribute to the team?" I picked this up very early. What he was doing was listening, motivating, and enabling us to play our best, which ultimately enhanced our careers.

Something that most people never knew about Red was that he liked to do what I call "little acts of kindness." Often, though, you wouldn't see it coming. He never broadcasted it or boasted, "This is what I'm doing for you"—he just did it. In the early 1960s, one of our forwards, Jim Loscutoff, injured his knee so badly, it required two operations. When Loscy came back to practice after his second surgery, he was naturally scared to test the knee. The doctors had told Red that if Loscy didn't rigorously break up the scar tissue in his knee, his career would be over. But Loscy wouldn't extend himself to work it that hard because it was so painful and he was scared he'd hurt it again. So Red developed a little workout routine he called "Fetch." He stood at the foul line with his back to Loscy, who stood under the basket, and Red threw the ball to either side, yelling, "It better not bounce twice!" So Loscy had to run hard and dive for every ball.

This went on for a couple of weeks. At first, witnessing these grueling sessions, I didn't understand it. Red seemed uncharacteristically cruel because Loscy was so frightened of reinjuring the knee. Plus, every time he dove, he got a painful "strawberry" burn on the wood floor. Loscy started to hate Red for this torture. He said, "When I retire, I swear I will come back here and strangle him for this!" Then one day the scar tissue had finally dissolved, along with the pain and fear, and that extended Loscy's career. So Red's "cruel" Fetch sessions were really brilliant little acts of kindness in disguise.

From the first, Red's method of coaching me was by observation, listening, and conversation. He perceived quickly that I was a very private person, a man of few words, and sensitive to how people talked to me. So when we sat down to talk, there was nobody else in the room. It was always on a personal, one-on-one basis, which I appreciated very much as a sign of respect. He didn't convey that he had authority over me, so there was none of the coercion I was used to, like "This is the way I want you to play" or "I'm the coach and you'll do what I say." In fact, in all our years together, he never once ordered me to do anything. His attitude was: "How can we do this together?" He approached it by asking himself, "Russell is a good ballplayer. What can I do to help him win games?" That would have been sufficient. But he went further: "This is a man I like and respect."

Once we started learning about each other as player and coach, he'd ask me basketball questions. That was how he gauged my psychology, my temperament, what drove me, and what I needed to be at my best. He never actually sat down with me and said, "Let's talk about your philosophy of the game." But he was very careful to keep all our conversations horizontal—"What do you think of this play? You think this'll work?" We never really talked as coach and player. We always talked to each other as men. It was very important to him, and he understood that it was very important to me.

You could say we treated each other as equals, but I shy away from the words "equal" and "equality." Sometimes, the implication is that one person who is somehow superior tolerates a "lesser" person. It almost becomes smug: "How liberal of me to accept you as an equal!" That actually suggests *in*equality, because it doesn't acknowledge the other person's humanity. In my entire relationship with Red, there was never even a hint of him treating me patronizingly as an "equal," as if it was a special favor that was his to dispense. So instead of saying we worked together as equals, I always called us "co-workers." We had a job to do on the Celtics, and as part of this job we each had a different assignment—but we needed each other to get it done right.

Even as co-workers, we didn't suddenly become friends one day. It took thirteen years of working together, thoughtfully and deliberately, doing something we both wanted to accomplish. Over that time, we learned that we shared core values, as professionals and as men, and that we could give and accept help from each other. It wasn't anything we analyzed or discussed. We just developed a "feel" for each other, and let things take their course.

Here's a subtle, early illustration. One afternoon my first year, Red and I came out of Boston Garden together after a practice. An excited guy hurried up to Red and said, "You're Red Auerbach! The coach of the Celtics! I'm a big fan! I think you're a great coach! Would you sign this?" He handed Red a Celtics program, and Red signed it. Then the guy handed it to me. "Here," he said. "You sign it too." I snarled back, "Bullshit! You can't talk to me like that!" And I walked away. My reaction was from my upbringing: *I will not allow anyone to impose his will on me.*

Now, another coach might have been shocked and critical of one of his players snapping rudely like that at a fan, especially in public. Not Red. He was completely at ease with it, and there were no repercussions. What that told me was, "Red is comfortable in his own skin. Just like I am in mine." We never discussed that incident. But Red liked to talk about it on his speech circuit every year because it always got a laugh. I wasn't used to that kind of level-headedness in a coach. I thought it might be a sign that he accepted me on my own terms without really knowing me yet.

In our profession, we were both passionate, driven, and single-minded. Away from it, we were standoffish by nature, not terrifically sociable. Neither of us bonded easily with others. So, since we had established our bond through basketball, we spent very little time together off court, and rarely talked about social issues. We felt they had nothing to do with us. He had a family and I had a family. He had a cadre of friends and so did I—and they were different people. One of Red's biggest passions was gin rummy. We probably played a thousand gin games after I became team captain, but I never played in his weekly games at his country club.

During our fifty years of friendship, we didn't know much about each other's private life. We never discussed anything we didn't want the other one to know. We didn't ask personal questions to satisfy our curiosity, and we never probed sensitive personal matters. He didn't tell me how his parents raised him in New York, and there were plenty of things about how I grew up I never shared with him. Neither of us knew if the other was a Republican or a Democrat. I didn't know if he went to synagogue, and he didn't know if I went to church. To most people that might sound strange, but for us, it was routine. That was how we were, and we liked it that way.

A big part of our foundation was that we didn't want anything from each other *except* friendship. So it wasn't necessary to share everything about ourselves. It was more as if we were saying, "You're here now and that's sufficient." We both knew what was important, in basketball and life, and what was irrelevant. That was how we overcame our differences.

Another quality I admired in Red—and this is a critical aspect of how to be a man and how to be a friend—was that he knew how to listen. If I told him something that was important to me, let's say, as a player, he would not only listen carefully but also absorb it and then act on it to somehow help me or the team. In fact, he was the best listener I have ever encountered. It was the secret of his success: *great ears.* Listening was how he first discovered that we perceived most things the same way. Another key element was that everything Red did in relation to me was thoughtful. Normally, friends slip up on occasion and say or do something uncivil or rude. I can't recall Red ever saying or doing anything uncivil or rude to me. He always behaved sensitively and carefully with me because he valued my respect.

Real friendships have ups and downs. Over time, there will be things the other person does that you don't like. But they're rolling hills, not mountains. You just say to yourself, "I wouldn't want it any other way because that's the way he is." The other person shouldn't have to compromise who he is, or change a thing about himself, to be your friend. You cannot base a friendship on illusions. I express my friendship equation this way: My ambition as your friend is that my friendship has a positive influence on your quality of life. If I am able to accomplish that, it will enhance my quality of life.

Another important ingredient in our early relationship that allowed us to click was discovering that our main common interest was our work ethic. We were both result-oriented— our egos were satisfied by doing good work. I was there to achieve something and I knew what it was. He was there to achieve something and he knew what it was. Fortunately for both of us, we recognized that our ultimate professional agendas were identical: *to win championships*. Red appreciated my work ethic, which came from my father. When I was about six, my dad sat me down and told me, "Son, I don't know what you'll be when you grow up. But here's what I want you to think about. When you take a job, if they pay you two dollars a day, give them three dollars worth of work. The reason is, if they're paying you two and you're giving them three, you're more valuable to them than they are to you. And if you do that all the time, you can look any man in the eye and tell him to go straight to hell. Because you worked, he paid." That was my ethic the day I showed up to work for Red Auerbach and the Celtics.

One of the pivotal moments in my career, and my relationship with Red, was the twelfth game I played my rookie year, in St. Louis. The first eleven games I'd been coming off the bench. This was my first game as a starter. Before the game, Red asked Cousy, our captain then, "What do you think, Cooz?" Cousy said, "I got Slater Martin guarding me. I can take him down in the blocks and I'll either get him in foul trouble or get a lot of easy shots." I was just a rookie, but I didn't think much of this idea. If you put your point guard down in the blocks where the center normally plays, everybody else is standing around out of position—that's really bad basketball. But Cousy went down in the blocks, and then I was out of position standing around while Slater Martin beat him up. The referees didn't call any fouls, either, so we were getting the crap kicked out of us. Next huddle, Bill Sharman wanted to go down in the blocks with his man. I was out of position again, standing around, really getting pissed because Sharman got the same results.

By the second quarter, we were about 18 points down, so Red called another time-out. The Celtics never sat down during time-outs in those days; we always stood together in our huddle. But I walked over to the bench and sat down. Everybody looked over at me until Red said, "Russell! Why aren't you in the huddle?"

I said, "I play center. Everybody else is playing center tonight. I don't need to be in the huddle to know how to get out of their way."

Now, in those days, rookies didn't talk to the referees, and they didn't talk back to coaches. Rookies weren't considered people yet. But Red thought about it a few seconds and said, "Okay, nobody plays center but Russell." Just like that. It caught me completely by surprise. There wasn't another coach in the league who would have taken that tack. Any other coach would have said, "Get your ass off the bench and get back to the locker room and get dressed! I'm the coach of this team!"

But now Red was in a very tough position. He had nine other players standing there, listening to this exchange between him and me. How would they respond to his acquiescing like that? Would it undermine his authority with the team? He had to take all that into consideration, right on the spot. His response told me that Red had decided, "Russell is the horse I'm riding." It was a critical moment for me—it started building our relationship. What if Red had considered me insolent, or confrontational, or disruptive to his authority? It very easily could have finished my Celtics career before it started. I could have picked up the reputation of a troublemaker and been moved around from team to team.

But Red's decision wasn't about his ego or mine. Both of us had the same kind of ego—if his team wins a championship, he's a great coach, and if my team wins a championship, I'm a great player. He didn't need to prove he was the coach. He didn't need to change me to get my best. His only thought was, "How can I help this team to win?" That never changed.

For the rest of that road trip, I played center all the time. That sent me another message about Red: he appreciated my perspective and my value to the team. It stimulated me to reciprocate.

Nobody knew it, but I had been assiduously studying all my teammates' tendencies, to learn their strengths and weaknesses so that I could help them all improve their games. On offense, I utilized my passing ability to distribute the ball to everyone exactly where they needed it to put up their best shots. The better my passing, the more we won. However, I was so effective at it, nobody was passing *me* the ball anymore. It annoyed me, but I never brought it up. I didn't have to—Red saw the problem right away. He realized that he had designed plays for everyone on the team except me. He knew also that, on every team, there was a hierarchy of who gets to shoot the most. You have a Number One shooter, a Number Two, a Number Three; this guy gets so many shots, and this guy, and this guy. The rest of the guys get the leftovers. On our team, since Cousy, Sharman, and Heinsohn had done most of the shooting before I arrived, Red knew they would probably put me in the leftover bin. He also realized—as I did—that for us to be a championship team, I had to be a major part of all the action on the floor.

So, at the first practice after we got home from the St. Louis trip, Red announced, "I'm putting in another play." He figured, "We've got this center and he's pretty good. How do we take advantage of that?" The equation he calculated turned into a play specifically for me to shoot. We called it the Six, for my uniform number. But instead of forcing it on me, he collaborated with me as my co-worker. "Okay, Russ," he said. "You're the center. We need you to shoot some more. Will this play work for you?" He walked us through it and we put it into our game plan, and I started getting more shots.

I had played three years in college, and then on the Olympic team, and there was never a single play for me to shoot on either team. No play where you'd say, "Okay, let's give Russell a shot." Did not exist. Nobody ever said, "Get the ball to the big guy." So everyone just figured, "They don't run plays for him. He must not be able to shoot." Now I had a coach who actually *wanted* me to shoot. The really interesting thing was that it was not a "reward" to me, or an acknowledgment that this is what we needed to do more often because I was the new guy. It was about Red's profound understanding of human psychology, especially in terms of how such diverse, high-achievement men could work together most effectively.

Red knew that, psychologically, all players needed to shoot in every game, whether they could shoot well or not. Some needed twenty shots a game, some needed two. He watched us closely—our focus, our body language—looking for any clue that someone might be thinking, "I need my shot right now." Whenever he sensed that, if nobody else did, he'd call a play from the bench to get that guy his shot. I picked up on this my rookie year, and I marveled at his perception. I knew what he was doing because I saw a lot of the same things.

For example, whenever Bill Sharman reached his "I need to shoot" threshold, his shooting hand would start to shake, as if he were in withdrawal from a drug. No one else saw it, including Red. So I went over to Sharman and said, "The next time they call my play, I want you to go over there." When my play got called, he went to that spot, and I passed him the ball, and he made the shot. After that, his hand stopped shaking.

Every game after the St. Louis trip, Red reinforced the new approach until it became part of the team's culture. If nobody was getting me the ball to shoot for a while, Red would yell from the bench, "Six!" Or in the huddle, he'd say something like, "Okay, Russell is busting his ass getting us boards and shots. Let's get the big guy some shots." I was aware that he was constantly evaluating each of us in every game, looking for any little way to help us win. I was the new element, so he was still calculating my equation. Here he had this tall, skinny kid who didn't play like other centers, yet he could play the position strong enough so that we no longer had a doughnut in the middle against other big centers—we had some strength in there too.

So now Red came to every practice calculating, "Okay, let's find out what this kid's strengths are." That's when Red and the team began what I called our "period of negotiations." It worked like this. At a practice, Red would ask me, "Russ, what do *you* want to do now with the Six?" That was a first—

a coach asking me what I wanted to do. It was also intriguing how he sensed that I wanted to vary the options on that play. It was as if we had a telephone line between just us, without any static and always on the same wavelength. I said, "Well, I would like to have guys cutting off me." Red thought it over and said, "Okay, let's get Russ some cutters." We set up the play with a forward cutting right by me on one side and a guard following behind him on my other side in an X pattern, like scissors. Red said, "Okay, Russ, where do you prefer the first cutter go through?" Again, a collaboration: not giving an order, just asking a question and leaving it to us to decide.

I loved how we did it. We had only six main plays. But by constantly engaging our creativity, Red engineered the system such that all five guys had something integral to do on every play. That way, the entire team was invested in every play. He stimulated a tangible spirit, pride, and team chemistry, and created so many avenues for everyone to contribute, it was exhilarating to be part of that flow. For example, about ten games into using the scissors play, my cutters realized that if they got a step ahead of their defenders, they'd get an easy layup because I was a great passer. Talk about being in sync with the way your coach thinks! Red and I were just getting acquainted and we were already thinking like twins.

None of my previous coaches had recognized my passing skills. Red saw them instantly. But instead of acknowledging this to me, or telling anyone else, "Russell's a great passer," he found subtle ways to tailor the offense to those skills. Eventually, a lot of our plays depended on my passes. Working together this way, we developed variations off those six basic plays until we had maybe fifty different options. Red engineered most of them to utilize my skills, an enormous sign of respect: *Russell is the horse I'm riding.* It reinforced for me that Red understood my capabilities the same way I did. In college, I never knew what the coaches wanted, so this was night and day. For the first time, I had a coach who not only appreciated all my capabilities but who invented a framework to unleash them—*for the good of the team.* That was exactly the responsibility I craved because the team game was the game I wanted to play.

Red continued to talk to me, listen, and observe me on the court, and he kept seeing things I could do that nobody else saw. Then he'd put them into the system: "Okay, Russ, here's another idea. Will this play work for you?" I might say, "I'll make some adjustments on it. But I can work that play, yeah." He started to respect my abilities and he tried to help me, and the team, to capitalize on every one of them.

To me, the key to this approach was that I did not lead him to that conclusion—we walked through it together. He started to trust my instincts and intelligence, the way I was starting to trust his. Simultaneously, our friendship started to emerge, although it was unspoken. I showed my apprecia-

tion by continuing his example and turning our play variations into something productive for *the team*. Sharman came
over to me at practice one day and said, "Russ, I'd rather take
a jump shot on the Six tonight. What I'll do is, I'll start to
cut past you but then I'll stop and come back a step. That'll
knock the defender off. He'll continue on through and you
can screen him off so he can't get back to guard me, and
I'll shoot. Okay?" If it was okay with me, it was okay with
Red—provided it worked in the game. I began assuming the
responsibility to make sure that it did.

In that kind of mature, fluid, collaborative atmosphere,
we kept developing our variations and creative flow. This
became one of our team's trademarks—you couldn't tell
which play was coming because the setup looked the same
for every option. Opponents found out abruptly that when
we set up, say, the Six, I was not locked in to just taking a
shot anymore. I could switch to the scissors option, if that's
what the defense showed us. I could pick any option off the
Six that I thought would work at the time, so they never
knew how to defend me. Red's mathematics again: *complex
equations simplified*.

This atmosphere was almost entirely Red's creation. It
was about his natural openness, powers of observation, psychology, mathematical genius, and willingness to allow his
players to improve their own games, as long as it improved
the team's chances of winning. This unique method started

to evolve immediately after that first road trip when I had told Red, "I don't need to be in the huddle." Later, I used to kid him about that. I said, "Red, I'm your Mephistopheles. The second you said, 'Nobody plays center but Russell,' you sold your soul to me!"

I was learning that Red was not only someone I could respect and trust, but that we might one day be friends. We were men working together as men. Just like in a friendship, acceptance isn't a gift you exchange. It comes out of respect—and you're not being "generous" for offering it. Instead, you feel, *That's the way it's supposed to be done.*

Chapter 4

||||||||||||||||||||||||||||||||||||||

My Father's Son

After a practice one day my rookie year, Red said, "Come in early tomorrow. I want to talk to you." The next day, I showed up early and we sat in the Boston Garden stands while the workers laid down the parquet floor. That was a whole process—converting a hockey rink into a basketball court—that I remember vividly. If I think about it today, I associate that image with special private talks between Red and me. It was an occasional ritual I looked forward to—talking basketball with Red while watching the conversion going on below.

This time, Red said, "I have to confess. I can't teach you anything. I don't know what the hell you're doing out there—nobody does. I just know it works. So I'm not going to mess with it."

I said, "I appreciate that."

"Listen, Russ," he said. "I want to ask you something."

"Uh-huh. What is it?"

"Do you know how good you are?"

"Yes!" I said, smiling. "I do."

"You're the best player playing basketball."

"I know that, Red."

"Okay. I just wanted you to know *I* know it too."

I was not bragging or being arrogant. I was confident because I knew what I could do. My attitude when I arrived in Boston was: "I am the best basketball player on the planet. Every game we play, I am, and I will be, the best player in the building." I believed that and I played like I believed it. But until this little chat, I wasn't absolutely sure that Red knew what I could do. Because he was right: nobody else knew. Until then, the Boston press was telling the fans that Cousy carried the team. Now, Cooz was a great player, but he did not carry that team. Red was saying he knew better.

What neither Red nor anyone else knew then was how prepared I really was my rookie year. My competitive nature only started to manifest itself in high school, and then on a California High School All-Stars tour of the northwestern United States and Canada. On that tour, I was like a sponge, soaking up basketball for the first time like I was obsessed. At first, I studied every player's moves and replayed them in my mind, over and over, picturing myself doing those moves. I had it down to their actual footwork—which foot they moved first on which play, which one next, and so on. Then I realized that I didn't want to do those moves myself so much as I wanted to learn how to defend against them. So I started practicing all this footwork in the mirror at night in order to teach myself the opposite footwork I'd need to defend it. By the time I got to the pros, I had scouted every player in the league to the nth degree. I knew what they did, when they did it, and how they did it.

After that tour, I remember telling my father, "I can play now!" I had found myself as a person—I would no longer be intimidated by anyone—and an athlete. All of a sudden, I wanted to find the best players in the world to play against, to prove to myself that I was as good as I thought I could be. I cannot overstate how important this was to me then. It was the strongest sense of self I'd ever known. I had all this new self-confidence, and, as my father had taught me, I had a plan. That plan was to prepare myself in college to become a champion player in the NBA. All I needed was the right environment and support. I didn't know I would find both on Red Auerbach's team.

In the 1950s there was a sports magazine called *Dell*, and I used to read the basketball issues to learn everything I could about NBA players. Most of the guys I read about, I never saw play. But I wished I could play against them. I studied those magazines and memorized every tidbit about each player: "He likes to use the backboard." "He has a sneaky right-hand hook." "He won't dribble left." I studied the centers especially because, of course, I wanted to measure myself against them. When I got to the Celtics, I not only knew the strengths of every player in the league, including my own teammates, but also their weaknesses.

All this scouting research helped me to determine which of my skills to use against which player, and when. It wasn't to make the All-Star team or to win a trophy as "Best This" or "Best That." I didn't care much about individual honors or hardware. It was to help improve my team's chances of winning. It didn't take Red long to perceive what I was trying to do. Once I became a starter, I was obsessed with finding new ways to improve on what I already knew how to do. At my best, I felt like I was playing inside a box where nothing else existed. Red figured that out really early because that had been *his* process when he was a player in college.

When I was in that box, I often slipped into what I called "my frenzy." It was a heightened intensity, a metaphysical state of synchronicity between my mind and body, and the game. I'd be so immersed in studying all the minute mechanics of the flow, there were periods when everything felt like it was happening in slow motion. The whole building went silent, and I could hear every bounce of the ball and see its individual grains. I'd tell myself, "This is it. I'm totally immersed. Now, which of my skills do I use? When do I use them? How do I use them?"

I never knew what my statistics were and I didn't care. Even in my private frenzy, I didn't feel it was just about Bill Russell.

I felt it was about creating an atmosphere in which my team couldn't lose. Red immediately understood this and trusted it. This was a guy who was result-oriented. And although we weren't friends yet, that's what friends do: they understand each other, first of all, and then they learn to trust each other. Very early on, Red and I developed an incredibly high level of trust. My previous coaches had said and done things that were not in my best interest, in order to make themselves look good. I knew that Red would never do anything like that. Even though he was my coach, there was never a feeling that he was wielding authority over me. We were working together as parts of the same unit. That was another reason why this friendship was so unusual. Red didn't feel he had to give me orders or keep me in line or steer me in a certain direction. And I never had to worry about picking up the newspaper and reading that Red said something like "This is what I told Russell to do." Whereas at USF, after I had that great game at San Jose my coach told the press, "No one player was outstanding," Red couldn't conceive of anything like that. He told the press many times, in fact, "The big guy came through for us again." That was all he *had* to say. In the unspoken language of male friendship, that light remark carried a heavy payload.

Of course, Red and I didn't always agree. But when we disagreed, usually it would take only a brief conversation to get the situation squared away. If he did something that I didn't like, I'd say, "Well, I don't particularly like that." He might say, "You know I'm not a frivolous person. I never do anything without giving it serious thought. That's the conclusion I came to." I'd say, "Okay." He did the same with me. In our first years working together, we both took the time to learn about each other, and to trust what we learned. I learned that he would never ask me to compromise my integrity. I also learned that he would never say or do anything that was not in my best interest, to benefit himself. I knew that he would never lie to me about anything—he always kept his word. Gradually, our friendship was built on this cornerstone of trust. At some point, if either of us asked a question, and there was a yes-or-no answer, either answer was acceptable. We both were willing to accept the other person as is. And we both made a concerted effort to understand one another, without shortchanging or compromising our principles in any way. *It is far more important to understand than to be understood.* Without ever discussing it, this became an organizing principle throughout our professional relationship.

Our friendship didn't happen instantly—it had to evolve. There were other ingredients in the mix. For example, in considering a friendship, I ask myself, "Is this a good person?" That's something you discover over time as you develop a feel for each other's core values. All the decisions Red made as my coach I understood clearly. However, when we were first learning about each other, we didn't share our perspectives and decisions on social issues—that wasn't part of the deal. So there were things we didn't know about each other off the court. For example, I was very concerned with civil rights issues, but I can't recall ever discussing this with Red. He knew I was a serious person, and that when people are serious about certain things, you don't push certain buttons. Yet in the late 1950s and early '60s, a few of them got pushed.

My first year as a Celtic, 1956–57, Red mentioned to me that he grew up in Williamsburg, a rough ethnic section of Brooklyn around the bridge that bears its name. He didn't offer much detail; he just said, "It was tougher than hell growing up in Williamsburg. There was a lot of prejudice against Jews. *I'm* a Jew." Not bragging and not complaining—just explaining.

I said, "Well, I don't know what that means. What *is* a Jew? Is it a religion? Is it a culture? A tribe?" At that time, I had heard the word on only a few occasions. For instance, at college, I heard about a popular booster called "The Jewish Don." That was about it.

Red said, "Russell. A Jew is a *Jew*!"

In all the years we knew each other, Red never did elaborate on that. I thought it was amusing. Sometimes during a game, maybe when Red was giving the refs the business, I'd walk by him and mutter, just loud enough for him to hear,

"Auerbach. A Jew is *a Jew*!" Once, I got a wink back for my trouble, and then he resumed dissecting the refs.

One time when we were discussing his definition of a Jew, Red told me a story about something that happened before I joined the Celtics. One day, he said, when he went into Walter Brown's office to ask him something, he found Walter listening on the phone, looking pale and appalled. Red said, "Walter, what's wrong with you?" Walter said, "Listen to this," and handed the receiver to Red. On the other end, some guy was ranting, "The reason the Celtics will never be any good is because they're a bunch of anti-Semites!" He was referring to a young Jewish prospect from Maine whom Red had just cut at a tryout. This guy on the phone kept raving about the kid's virtues, thinking he was talking to Walter. Suddenly, Red exploded into the receiver, "This is Red Auerbach! Listen, you fuckin' *Heeb*! *I'm* a Jew, goddamn it! *I* cut that kid—he couldn't fuckin' *play*! Walter Brown is one of the finest human beings on the whole goddamn planet! And you're talking this shit to *him*? If I find out who you are, I'll come over there and kick your ass!"

Several things struck me at once. It was the first time I'd witnessed one Jew teeing off on another Jew. That intrigued me. I had caught a lot of stereotypical slurs from white people over the years. But sometimes, I even caught grief from other blacks. For instance, I was in a club in Boston one night when a black guy approached me and said, "Bill Russell, you think you're quite something. But I want you to know, I'm a fourth-generation Bostonian. And we will *never* accept you." I said, "Well, I guess I'll have to learn to live with that." I realized that "acceptance" was his little red wagon. So that Jewish guy's rant on Walter Brown's phone told me that we weren't the only ones catching this flak from one of our own.

Another thing about Red's story that struck me was that his diatribe against his fellow Jew was, mainly, an act of loyalty to Walter Brown. Maybe more important, it hinted that Red's Jewishness and my blackness might not be a problem for us later on.

My first few years with Red, I carried my father's Bag Factory caution, "Never let your guard down," from that time he'd asked for a raise but didn't get it. So I carefully observed how Red treated each of his players. At that time, my perception was that most white coaches treated black players differently. My impression of their attitude was, basically, "If I play this guy, he should go through a brick wall for me because I gave him a chance." As if it were a favor from a white man to a black man, as opposed to playing the best men on the team, period. These coaches didn't bother either praising or consoling their black players. They just made judgments based on whether the black players were grateful for the chance to play.

I had seen this in college, so I knew it was happening to black players in the NBA. They told me stories about coaches who singled them out—yelling, criticizing them constantly—

based only on preconceived notions, stereotypes, and prejudices. I would not put up with that. If I had thought that Red
was yelling at me about things he didn't yell at his white players for, I would have been deeply offended, and let him know
it. It probably would have destroyed my trust and respect. But
it never happened. He always behaved in a civil, respectful,
evenhanded fashion toward me and everyone else.

However, my first year, I suspected that my being the
only black guy on the team might have made it easier for
Red, because that eliminated the "you people" mindset. It
was just "me," singular—he only had to deal with me one-
on-one, man to man, as opposed to me as part of a group of
"The Black Players." But very shortly I realized that Red
was a genuine good guy who didn't care about your race or
the color of your skin at all. All he cared about was whether
you were helping the team to win its games.

In those first years, there was one uneasy moment, during
my second season, 1957–58. In the spring of 1957 we had won
the Celtics' first NBA championship. The following fall, after
the '57 draft, Red called me up and announced, "I drafted
Sam Jones out of North Carolina Central University. How
about this guy, Russ? You think he can go good for us?"

I said, "Who the hell is Sam Jones?"

"He's a *Schvartzer!*" This was colloquial Yiddish for a black
person. "I thought you'd know about him." What flashed in my
mind was the old racial stereotype: *All black people know each other.*

I said, "Listen, Red. I don't know *all* of them!"

He laughed, which took the edge off. Yet it stayed with me a while. I was raw on racial issues then, and Red and I were still scouting each other as people. His remark may have been naïve, or streetwise, or maybe just a needle. Since by then I'd known Red over a year, I finally accepted it as inadvertent, and that was it, and we moved on. After that, we shared other potentially tense moments, but eventually, as we grew friendlier, we could say anything to each other, without hesitation, and immediately understand what we both meant. So, as it turned out, this little event triggered an understanding between us that brought us closer to a deeper friendship later on.

At some point, Red and I recognized that while he and I came from different tribes, he was the same within his tribe as I was within mine. That was our common ground as men. It wasn't like a Jew being friendly to a black man or a black man being friendly to a Jew. It was, mutually, *This person, from this tribe, is to be respected.*

But in those early times together, there were still some tests. Take 1958. That season, Red scheduled a regular-season game in segregated Charlotte, North Carolina. When I found out, I asked him about it, and he said, "Don't worry about it. Few years ago, we played a game there with Chuck Cooper. It worked out fine." He was referring to a black player the

Celtics drafted out of Duquesne in 1950, officially integrating the NBA. I was a little uneasy, but I figured he knew what he was talking about. However, what Red didn't tell me was that after they had played that game, Cooper was consigned to a segregated hotel. He was so upset about that prospect, he took a late sleeper out of town instead. Cooper's teammate Bob Cousy decided to accompany him—a sensitive gesture for which I always respected him.

Red wasn't with us on the plane to Charlotte, so it was our trainer who informed Sam Jones and me, "By the way, you fellas are staying at the Such-and-Such hotel." Subtext again! I said, "Hold on. What's this 'You fellas' shit?" My mind jumped back to an incident in college, when USF flew us to New Orleans for a game with Loyola of the South, a sister school. It would be our first game in a segregated town. On the flight over, our trainer informed the three blacks on our team that we'd be staying in the dorms at Xavier University, while everyone else roomed at a hotel in town. So the only time we saw the rest of the team on that trip was at the arena for practices and the game. But, okay, they were doing this little social experiment, which was really what it was about. Many years later, I told my father this story. I said, "That really annoyed me because they never once asked us what we felt about it." He said, "Well, they asked me." I said, "They *what*?" He'd never told me about it, but apparently, instead of asking me if those rooming arrangements were okay, the school asked my father behind my back. I harbored that outrage for years, until I finally let it drain. Now, in 1958, I was facing the same situation.

We found out soon that, sure enough, we were staying in a segregated hotel. Sam and I discussed it, and we figured that while Red was busy booking all the games for the season, he didn't pay this one much attention, and he just failed to anticipate these kind of complications. So, as professionals, we decided we owed it to our teammates to play the game. But afterward, I phoned Red at his hotel on the other side of town. I said, "Why are we in this goddamn dump?"

He said, "You know how they are down here, Russ. It's not that bad, is it?"

"It is to *me*!"

"Well, I can't control these things. I can't help how these people are."

I was not upset. I spoke softly but firmly, "That's bullshit, Red. You booked this game. You booked this place for us to stay. You know how these people are. So, what you did was you put us in harm's way."

He said, "Remember, I'm a *Jew*. I got to be careful down here too. I got to deal with the same shit."

I said, "Oh yeah? What hotel are *you* staying at?" He had no response—what could he say?

I wasn't angry with Red or anyone else. I just thought, "This is wrong. I can't think of any logical reason I should have to put up with it." It wasn't about pride or ego or creating conflict. It was about principle, trust, and respect. I said, "Red, I am not a spokesman for the black players. I'm speaking for myself. It's okay this time. But it'll *never* happen again." I left it right there. I think this exchange conveyed a lot about me to Red. I had told him what it was about for me, and that I was not concerned with how he assessed it or what he would do with it. I was not testing him at all—I was making a point that had to be heard. And he heard it, and we moved on.

Afterward, I ran the whole thing through my mind. I felt that, in regard to Red and me, it was part of the process of learning to accept and trust each other. I understood, first of all, that in this particular situation, Jews were generally not welcome there either. Still, blacks were treated differently from Jews or anyone else who wasn't black. As I said before, I learned from my father that minorities have to understand the majority to survive, but not the other way around. I didn't think this was Red's attitude—he didn't treat people that way. So, what it came down to was this: First, it was an oversight. I figured, "He books all the games. When he booked this one, he never thought about the hotel situation because it hadn't happened before." Second, up until then, Red had demonstrated to me that he made professional decisions based on logic, common sense, and what was best for the team rather than himself. Third, I knew that Red had come up against prejudice as a Jew in Brooklyn, so he'd walked a mile in my shoes. He didn't want his own sensitivities dismissed, and so far, he had never dismissed mine.

Something else about Red's character came to mind. My rookie year, a reporter who thought I was hard to coach asked Red, "How do you handle that Russell?" and Red told him, "You handle mule teams. You don't handle men. You treat men with respect." That left an impression. So I decided it was better to understand than to be understood, and left it alone. However, I felt I also had made an important point to Red about *my* character, which was simply that I do not ask for understanding. I have never worked to be understood, or accepted, or liked. So no explanations are necessary. I care only about what you do. I never raised this with him because I would never explain, or justify, or defend myself to others. I just figured that the next time he made out a schedule, he wouldn't book us in that kind of hotel. And I was right. The next time we played down South, a couple of years later, the whole team stayed at the same hotel. Red had made the adjustment so it *didn't* happen again. It was in the past, and I didn't think about it anymore.

On another occasion in the early '60s, we were scheduled to play a special homecoming exhibition with the St. Louis Hawks in Lexington, Kentucky. It was to honor two former University of Kentucky All-Americans, our Frank Ramsey and the Hawks' Cliff Hagan. We had to fly all day on that trip, and we never had a chance to eat. So when we checked in to the team hotel, food was the first thing on our minds. As I remember it, I was leaving my room to go eat with K.C. Jones when Sam Jones and Satch Sanders stepped off the elevator. "Where you going?" Sam asked us.

K.C. said, "To get something to eat in the coffee shop."

"You can't eat down there," Sam said. "They don't serve colored people. Waitress wouldn't serve us."

"Sam," I said, "that's the best damn thing you ever said to me. I'm going home! I didn't want to be here in the first place." We talked things over a little bit, and then I went back to my room and called the airport to book myself a flight home. The other guys decided to leave also, and made their own arrangements.

Then I phoned Red. I said, "We're going home."

"What're you talking about?"

"Well, Sam and Satch went down to the restaurant and they refused them service. So the black players on our team aren't playing. I don't know about the rest of the team—that's *their* choice. If they play the game, I am completely okay with it. I'm speaking only for myself now: this is *my* choice. I got a reservation for the seven o'clock flight."

It was clear that I wanted the team to play the game without us. It was better for two reasons. One was that Red had contracted to play that game, and I wanted him to meet his obligation. Number 2, if all the white guys played, it would only increase the impact of our message.

Red said, "Well, hold on, wait a minute. Let me find out what the hell's going on."

A few minutes later, he called back to say, "I took care of it. Everything's okay. I got things squared away."

"Oh yeah? How's that?"

"I talked to the owner of the hotel. He's also the manager, and he invited you guys to all have dinner with him in his private suite."

I said, "Red, I don't know *him*. And I'm not interested in sharing a meal with him. Who the hell does he think he is? I don't want to have dinner with *him*!"

"Okay, okay. Let me find out something else."

He called back again and said, "Well, he said that you guys can go down there to the restaurant and eat right now. And he swore to me that they'll never segregate again."

Now, from the owner's perspective as a member of another tribe—which I completely understood—this abstract, self-serving concept of "fair play" was an acceptable compromise. That was his tribe's frame of reference. But, for me and my tribe, there was no compromise available. So I said, "Red. Listen to me. I've heard all the arguments, and I am no longer willing to accept the status quo. There's nothing happening in Lexington tonight that will keep me from taking that plane at seven o'clock."

Silence. Finally, he asked, in a soft, civil tone, "That's the way you feel about it?"

"That's the way I feel about it."

"Okay," he said flatly. "I'll come out to the airport with you guys to make sure your tickets and luggage are all set." He accepted our decision—just like that. The guys knew implicitly that I liked all of my teammates, white and black. But this was something the black guys had to do for ourselves. We had to make a choice *for us*. Red had a choice to make too. He could have said, "You don't play, I'll fine you." But that would have meant siding with the people who'd refused us service. If he had done that, we never could have been friends.

He went with us to the airport instead. Red was not angry at all. He knew, by then, what kind of man I was. And I knew by then that he would never front this issue with me again, that he would never say to me directly something like, "Bill, you're my guy. I understand you perfectly. I'm with you guys all the way." If anyone tried to tell me something like that— "I'm not a racist, you know"—I wouldn't have heard him. *I care only about what you do.*

So, Red, K.C., Sam, Satch, and I went out to the airport, and we changed the tickets and left. The reason I didn't arrange a pregame press conference for all of us who were leaving was that I always tried to avoid becoming a victim. If I had announced to the press, "They wouldn't serve us in the restaurant," they would have written that the black players on the Celtics were refused service before the game—and that would have portrayed us as victims. Instead, the next day it came out in the papers exactly the way I wanted it to: "The Negro players did not play." Period. Our attitude was, "We've been offended, and it's about respect and principle, so we're going home"—and that's what we wanted to convey.

Once we left, Red returned to the arena and the team played the game without us. The people of Lexington were happy with that. They got to see the lily-white game they probably wanted anyway. It was an exhibition game, and that's about the franchise making money. To Red's credit, unlike what I had experienced in college, when we played down South, he never involved himself in social experiments: "Let's do this for the black guys" or "Let's do this for the Jewish guys" or "for the white guys." No such social engineering. That was one of the prime reasons I could trust and respect him. On the other hand, you don't want a friend who you have to nominate for sainthood because, in his reflection, you will see all your own character flaws.

There was something else at play that almost nobody else knew about me then. When I was ten, I read a black history book about a revolutionary Haitian dictator named Henri Christophe. It caught my attention that he'd built an enormous fort on a cliff in Haiti to defend his people against marauding white enslavers and European colonizers. His fort still stands today. What impressed me was that, in this book, the author said, "This fort is the only monument in the Western Hemisphere built by a black man." He didn't use the word "Negro." He used "*black man.*" That was the first time I heard that phrase, and I liked it.

But the author also said that, in those days, white Americans commonly believed that African slaves were better off living in America as slaves than as free men in Africa. I remember thinking, "How can you say that?" I couldn't accept it because I thought the credo of America was "Freedom!" I was only just beginning to understand that, for many Americans, it was more like "Freedom for *us*, not for you!" Instantly, Henri Christophe became one of my heroes. By the time I made my first trip to Africa when I was twenty-five, where I happened to meet some of the freedom leaders in what I called Black Africa, I was already referring to myself as black.

So when these racial incidents happened down South, my attitude was not that I was being treated badly—if you feel like that, you've made yourself a victim. My attitude was more like, "Okay. *That's* the way you want to do this? Cool!" In Kentucky, when Sam Jones said they wouldn't serve colored people in the restaurant, part of me detached immediately and I just thought, "Okay. It's time to eat. But I can't get any food here. So I'll go home. I know I can get a good meal there." It may sound odd, but at that moment in time, I had absolutely no sense of being oppressed. I felt like a hungry black man, anxious for home, not just another angry "Negro" victim.

There is a time-delayed postscript to this story. Two years ago, the head of the alumni association of Prairie View College, a black school outside Houston, held a couple of fundraisers. For one of them, he asked my friend, former baseball star Joe Morgan, and me to do a question-and-answer fundraising session with the alumni association. The night before, at dinner, the president told me, "Bill Russell. You don't know how much of an impact you've had on my life. A significant impact."

"How?"

"When you guys wouldn't play at Lexington, Kentucky, I was in school there. When you told the white people of Lexington, 'You can't just do anything you want to me,' they started looking at the black community a little differently."

"I didn't know that."

"I never thought I'd meet you. But all my adult life, I wanted to say thanks."

I said, "Well, all *I* was doing was being my father's son!"

When I said this, it summoned up a night, years ago, when my very dear friend Jim Brown called me at 3 A.M., which he still does a lot today, and said, "Couple weeks ago I met your father, and I was thinking about you, and I want to tell you something."

"What?"

"You ain't *shit*!"

I said, "Well, *I* knew that. How'd *you* find out?" I rolled out my big laugh.

He said, "All this time, I thought you were this exotic person who came out of nowhere. But after meeting your father, I realized: You're just *your father's son!*"

That was a tremendous compliment, because Jim understood what my father stood for, and what he meant to me in my life. Red got that too. In fact, I think that he would have been disappointed, at that stage of our relationship, if I had acted any other way.

After we got back to Boston, when Red told Walter Brown what happened, Walter held a public press conference and said, "I apologize to my players. I should have never let that happen." It was in all the papers the following day. Later, I acknowledged to Red that I appreciated his decision to come to the airport with us. I considered it a gesture of respect. I remember thinking of something my father taught me, "You must acknowledge and accept other tribes and never say 'My tribe can do this, so they're better than yours.'" I'm not sure, but I may have shared this with Red at that time.

We never revisited those incidents again. But I learned over time that I was right about Red. He did understand why I protested in Charlotte, and why we had to leave Lexington—he communicated that chiefly by not trying to impose his viewpoint on us and then making the necessary adjustments so that we never had to deal with that nonsense again. Pretty soon, it became clear to me that one of Red's best qualities was that he always assessed important situations so he would know how to react to them if they ever happened again. For himself—not to save mankind.

It was how he lived his life—he was open to improving himself the same way he strived to improve his team. To me, these were strong signs of the kind of man with whom, one day, I might forge a lasting friendship.

Chapter 5

What's Best for the Team

The early racial incidents down South crystallized some important intuitions both Red and I had about each other's character. We recognized that disagreements were not obstacles to our mutual respect, but rather stepping-stones to understanding.

Although I had started to accept and trust Red as a coach and a man, I was still in the process of integrating myself into the team and discovering what everyone else was all about, too. So, at that stage, my focus wasn't "How do Red and I become friends?" It was more like "How do we all learn to work together successfully as men?" To me, that process started with my relationship with Red. My journey of discovery of his true character continued with me learning more about his coaching methods, principles, and leadership.

. . . .

While most outside observers assume that the typical coach–player relationship is cut and dried, for Red and me it remained a complicated equation. Even if two people promise that they will be square with each other, the deepest friendships require more of both. You each have to know that the other is a good guy who will treat everyone else fairly, too. You can't have reservations about a friend's character, because it will eat at the friendship. I came to see that Red treated everyone the right way, but it took some time.

One thing I noticed about Red was that he didn't seem to have any preconceived notions. And if he had any deep-seated biases, they didn't affect his decisions. We all have common human frailties, biases, and preconceived notions. But, in spite of them, we have to make an effort to learn as much about each other as possible. Red and I did that. Our learning process was a two-way street, and we never did measure who learned more.

There was an occasional detour. Like the time at a pre-season practice my first year when I hastily insisted to Satch and K.C., "Red's a racist, you know."

K.C. looked shocked. He said, "What're you talking about?"

I said, "Okay, I'll show you. Watch this."

When Red came into the locker room, I asked him about a particular white player trying out for the team. Red said, "Yeah, he can really shoot."

I waited twenty minutes and then I asked him about a black player trying out, "What do you think of him?"

Red said, essentially, "He can't do this" and "He can't do that."

I went back to Satch and K.C. and said, "See? The way they evaluate white players is different from the way they evaluate black players. If they keep a white player around, it's because of what he can do. But when they evaluate a black player, it's about what he *can't* do."

We laugh about that today because once I learned more about Red, I realized that my own preconceived assumptions were wrong—there was a lot more to the situation than what I'd thought. For one thing, the two guys played different positions. In the black player's case, given Red's specific requirements for that position, this guy didn't have what he wanted, in terms of improving the team. As for the white player, all he could do was shoot, so Red cut him too, even though his shooting could have been a valuable asset. That incident only reminded me yet again: *It is far more important to understand than to be understood.*

The whole time that Red was with the Celtics, he never conveyed that his criteria for winning basketball games had anything to do with his social conscience, or some "liberal" philosophy, or anything like that. The only thing he cared about was what was best for the team. Here's one way to describe the atmosphere of integrity that Red created on the Celtics. In 1962, we had an all-black starting team for the first time. But it was two weeks before any of us realized it—and that was only because we saw it in a newspaper column. Even then, my only thought was, "Really?" And I never heard anyone else on the team say a word abóut it, including Red.

It was a routine A-B-C equation for us: *We're only interested in winning. This is our best team today. This is who we're putting on the floor.* Some people might call this a practical management philosophy, and it probably was. But you can't separate your perspective on life from the work you do. It's all part of the same view. The window through which we viewed the world was a little different from the window through which the world viewed us. When we existed in our Boston Celtics box, that box was all we cared about. Five black guys on the court at once? "Really?" We didn't notice. We wouldn't have cared if we did. It wasn't important.

But the writers pressed for a comment. Their little red wagon for the day was a fishing expedition for a momentous statement from Red. Something grandiose like, "We thought it was time to affirm that the Celtics believe that all men are created equal." But the one thing that Red hated more than a bully was a bullshitter. He told me almost the first day I joined the team, "I don't bullshit and I don't listen to bullshit." That registered because it might as well have been me talking. So he disappointed the press. He said, "I wasn't making any statement. I was trying to win games! Those were my five best guys. It was no big deal."

It *was* a big deal—to everyone else. We had just made history. "History," as Red would have said, "can kiss my ass in Macy's window." It wasn't as if we weren't "race-conscious"; all of us are race-conscious. A guy in Boston once said to me, "I don't notice whether someone's black or white." I said, "You better get some help. Talk to someone. Of course you notice!" In our society, there is so much emphasis put on our differences. We all carry around race consciousness and preconceived notions about others. The significant distinction is, if you allow these biases to command your actions, then you're a bigot.

Red didn't operate that way. I never felt that he was "politically correct" when he talked to me or anyone else. He was honest in the purest sense of the word. He spoke frankly but thoughtfully. He practiced a motto of mine: "Let the first thing out of your mouth be your second thought. It will greatly reduce the number of apologies you have to make, either to yourself or the person you're speaking to." Red was always just Red. And I have always been just me.

After I came aboard with Boston—where I was the only black player my rookie year and we won the championship—other teams started thinking, "Maybe we should take another look at these guys," and they started drafting more black players. But only so many. For years, into the 1960s, they enforced a quiet quota: every team but us had exactly three black players. How they came up with that number is anyone's guess. What wasn't a guess was the fact that whites made up the fan base then, and the white owners were scared to death of "offending" the base that paid their freight. The other part of that equation, of course, was outright prejudice: "We don't like black people."

In the middle of one of our win streaks, a Boston writer essentially said in his newspaper column, "The Celtics have too many black players and the fans won't stand for it." How dumb was that? The next day, he woke up even dumber, because he came to our practice to try to talk to me about his article. I told him, "Get the hell away from me. You out of your goddamn mind? I don't want to talk to you." He said, "Well, it's the truth, isn't it?" I said, "That's *your* truth. Leave me out of it. It doesn't have anything to do with me." In those days, reporters felt privileged, and smarter than us because the common opinion was that athletes were nothing more than "dumb jocks." Another example: My last year in Boston, 1969, Red and I marveled at a survey the Celtics took to see what the fans wanted for the team in the future. Over 50 percent said, "There are too many black players on the team." Too many black players on a team that won eight straight championships for the city of Boston?

One time, I called the league on it. I told the commissioner, "You're running a three-blacks-per-team quota. Not one, not two, not four—*three*! I do not accept this as an accident." He just said, "Well, I disagree." But under Walter and Red, the Celtics got ahead of that curve. I always liked to say that Red viewed players as silhouettes—race, color, tribe were irrelevant. For him, the only relevant thing was: *"What can this player do to help us win?"*

In Boston in the early 1960s, blacks weren't too welcome, especially me. Since I was outspoken about issues like civil rights, I took some shots in the press. When a magazine writer wrote about how he thought I hated my teammate Bob Cousy—which was completely untrue—Red said, "Boy, they're really trying to give you hell."

I said, "Doesn't matter. The press is irrelevant to me."

He said, "Did I ever tell you about what they tried to do to me when I didn't draft Cousy?" And he proceeded to launch into something that happened to him in 1950, his first year as Celtics coach. At that time, Cousy was the most popular college player in New England as an All-American local hero from Holy Cross. Everyone up there knew Cousy for his remarkable ball-handling skills. Red thought—wrongly— that Cousy was a little cocky. The Boston fans and press thought—also wrongly—that he was the greatest player on earth. Celtics owner Walter Brown absolutely loved Cousy too. Before the draft, he kept reminding Red how much the fans loved Cousy, and how great he'd be for our team. Plus, the league had the territorial draft back then, which made Cousy an even more obvious Boston pick. Except to Red. He had his own plan.

At the time, Red was focused on the NBA champion Minneapolis Lakers, and their big center, George Mikan. Red wanted a championship too, so he wanted a big presence in the middle to play against Mikan. There was a kid in the draft from Bowling Green named Charlie Share—seven feet tall, 280 pounds, with some skills. Red told me that his idea of a great team was one that could compete with the champs. "You don't build a team like that with little guards," he insisted. "You build it with strong big men."

So Red drafted Charley Share. "Russ," he told me, "the goddamn press went nuts. I'm a Jew—you'd've thought I was Judas Priest! They ate me alive!" Walter Brown was upset with Red, like everyone else. Even though Walter was a polite, mannerly gentleman, he gave Red a pretty good blast. Red just growled back, "Jesus Christ, Walter. You want me to win games or please the local yokels?" Red was not especially diplomatic.

It got worse. That same day, one of the newspaper hounds couldn't contain himself. He rushed up to Red and let him have it with double barrels of disgust: "You asshole! Don't you realize you've insulted everyone in New England by not drafting Cousy! What the hell's wrong with you? He's the best player in the country! Anyone with any brains knows that! Besides that, you're a Jew, and we don't like Jews either! And we're gonna run you out of town!"

When Red told me this story, I asked him, "Now, how did you handle *that*?" He took a long drag on his cigar and blew smoke in the air. "Oh," he said with a grin. "I'll just outlive the bastards."

I wasn't surprised at what that reporter said. I'd taken some ignorant blasts in the papers myself by then. And I knew, from how I grew up, that some people, in the security of their entitlement, think they can say or do anything they like, without repercussions. Then they're surprised when you take offense. Although Red and I never talked about that, we had both experienced it the same way.

It was ironic, too, because the next season Red ended up getting Cousy anyway. The Chicago Stags had drafted Cousy, but they folded after the 1949–1950 season. The NBA held a "dispersal draft" for its remaining teams to pick up Chicago's players by throwing the names in a hat. Cousy's name got picked by the Celtics out of Walter's old, brown fedora. The point is, Red didn't give a damn what other people thought of him. All he did was stay true to himself and try to do good work. Another trait that linked us.

Here's something interesting about how deeply I had absorbed Red's "I'll outlive the bastards" attitude of 1950. His will to outlive hate, ignorance, entitlement, and even evil— which echoed my own perspective—rushed into my mind one day fourteen years later in an awful place. As basketball grew more popular worldwide, in '64, at the request of a friend at the State Department, Red agreed to lead a goodwill tour behind the Iron Curtain. We went to Poland, Romania, Yugoslavia, and even Egypt. He took me, Cousy, K.C., and Heinsohn from the Celtics, and some All-Star players from other teams. Like me, Red loved adventure and travel and learning about new cultures, so he arranged for interesting cultural tours along the way. When we were in Poland, our guide asked us if we wanted to see Auschwitz, so we spent a whole day there. We saw the ovens, and mounds of hair and jewelry and shoes, and then we just walked around somberly.

Red looked solemn—it made a big impact on him. It had a strong impact on me, too. I thought, "This is the utmost demonstration of men's inhumanity to other men. This is a great evil." And based on what? People always say, "Well, Hitler was a madman." But he had help; even civilians were complicit. I could only wonder what Red was thinking and feeling. I didn't say a word to him the whole time we were there. I thought if I said something wrong, it might sound disrespectful or be uncomfortable for him—this place had more to do with his people than mine.

However, the place did summon thoughts about segregation in the United States. While that was different, it arose from the same ignorance and fear that triggers such cruelties. The philosopher Edmund Burke said, "All that is necessary for evil to triumph is that good men do nothing." That was what happened at these camps. Reflecting on that, I thought, "Sometimes all you can hope for is that you'll outlive the bastards."

Red and I did not need to speak to each other about the tribulations of our tribes. We did not need to express empathy. We were there for each other as men, not as representatives of our tribes. But it meant a lot to me that we could be friends, despite being from different tribes. Being at Auschwitz together that day helped to drive that home.

As my coach, Red Auerbach won nine championships, including a still unmatched eight straight. He won seven more titles with Celtics teams he built later as general manager. We were together for nearly all of his championships, so I had the pleasure of watching him work, with an advanced understanding of what he was doing and why. I saw how much went into making him so successful and making the Celtics the world's most successful team. He was a brilliant psychologist, strategist, innovator, and motivator, and the best coach you could play for.

Red was decisive, and I admired this quality in him. It was another little thread that we wove into a friendship. One early decision he made went a long way toward convincing me he was becoming a friend, and not just my coach. My rookie year, since I influenced our game so heavily on both offense and defense, Red played me the whole forty-eight minutes—nobody else played more than forty. But I went so hard every single game that, after a while, I lost my edge. I fought it off, so it wasn't obvious. If I wasn't hitting my shots, I intensified my rebounding and screening and defense, and I didn't think anyone noticed the difference. But someone did.

During practice one day at the Garden, I was sitting on the scorer's table, a little drained, and up strolled Red, looking concerned. He said, "Jesus Christ, Russ. You look like a piece of shit." Lovely greeting, but direct: he spoke his mind, so you always knew where he stood, and where you stood, too. "What the hell's wrong with you?"

"I'm tired."

"You're tired?"

"Yeah."

"Okay," he said. "Don't scrimmage today." Then he went back to the practice. From then on, once I got in shape, I'd come to practice and run through our plays with everyone else because that was something we needed to do together. But when it came time to scrimmage, I went up in the stands with a cup of tea and watched. That was great for me because I hated practice—I thought, "I already know exactly what I'm doing. Practice is only a drain on my energy." I found it remarkable that, in our short time together, Red not only perceived this about me—without my mentioning it—but found a solution to the problem.

Eventually, tea in the stands became my new routine, and I almost never scrimmaged again. Later, I heard from players whose coaches knew about this, and they all thought that Red was kissing my ass. With that narrow frame of reference—which was, essentially, a preconceived notion—they couldn't fathom it because they weren't capable of handling a situation like that with such clarity and wisdom. Some coaches today, when their team is going bad, call a midnight practice and work their players harder. That's counterproductive, like a punishment detail that only breeds resentment. Red's perspective on this was unique. He understood that it was tough enough to play NBA basketball, and that fatigue was a major factor. So why would you invite *more* fatigue? It became obvious to me that, where most coaches ran two layers deep, Red ran ten.

My second year—my first full year—Red expanded the Russell Fatigue equation. He told me, "I want to try something to get you some rest in every game." His solution, as usual, was simple but ingenious. He would play me the first ten minutes of the first quarter and rest me the last two minutes. Added to the three-minute break between quarters, that would give me a continuous five-minute blow. Then I'd come back and play the whole second quarter to halftime, have the halftime break, and play the entire second half. That way, Red could play me forty-six minutes a game—still a lot of minutes—but not wear me out. It was a huge help;

it reduced my fatigue and I felt more energetic. There is no doubt it helped us win more games.

Red kept calculating, though; somehow, he thought the equation wasn't perfected. My view at the time was that when I was in my frenzy to win, I created a groove where I played at a very high level. Inside that groove, it was as though I was on automatic—I was completely attuned to the flow of the game. But experience had taught me that even if you're getting some rest, if you try to sustain that groove too long, it morphs into a rut and you don't even know it. So you need to step back, step right, step left—do whatever you can to recover that groove. That was where Red's exceptional insights into people came in. He processed everything, from what you said and how you handled pressure to your practice habits and body language. After studying me intensely, he could always sense me drifting out of my groove and into a rut, sometimes before I sensed it myself.

Not long after we instituted the 46-minutes routine, he called me in to talk again. He said, "You know, Russ, I've been thinking about it. I'm playing you forty-six minutes a game, and then I turn around and wear you out all week in practice. Doesn't make sense." Red liked to use these two phrases: "Makes sense." and "Doesn't make sense." They were his basic problem-solving yardsticks.

Around the time of this talk, we had a rare four-day break coming up. So, at practice one day, Red told me, "I don't want to see you here for three or four days. Don't come to practice, don't call me, don't think about being here. I don't care where the hell you go or what you do. Get away and stay away and come back fresh. But remember this—you *owe* me one." That was his way of letting me go recharge my battery, something I really needed and appreciated. The attitude that he conveyed to the team was, essentially, "Everyone out there knows that Bill Russell has his back and that he's always ready to play when the game starts. So, if the big guy needs a rest, no big deal."

In this way, he was demonstrating that *he* had *my* back. In fact, he never once took anyone's side against me, and I never took anyone's side against him. That's a very important element in any relationship that you need to be able to count on. And we never had to say a word to each other about it; it just was. I thought, "He understands my professional demeanor. Here's a coach looking out for me for a change. Just like a friend." I perceived it as loyalty, and it carried into our friendship.

This was also Red at his motivator-master manager best, simplifying a complex equation. He sensed exactly when I needed to get away from basketball, and he knew I would never request it myself. But since I was so integral to the team's success, he knew that something had to be done. His genius was not in just finding *a* solution, or even a *good* solution. It was in finding the *perfect* solution. In this instance, he came up with a practical, though unconventional, one that not only served my best interests but also the team's. And the master psychologist in Red added a little interest clause that he knew would pay dividends for the team at another time. It turned out that what he meant by "You owe me one" was that, in some future game that he wanted to win really badly, he would expect me to make absolutely certain that we won. "Russ, you know you owe me one, right?" "Yup." "Well, I want it tonight." So I'd go extra hard that night, just as he'd calculated, and motivate the other guys, to make sure we won that game for him.

Starting then, several times each season, I'd vanish for days without telling anyone I was leaving, and no one would see me again until I showed up at practice. I knew it was a mystery that my teammates wondered about from time to time. Yet no one ever asked me about it. And as far as I knew, nobody asked Red, "Where the hell is Russell?" I never knew how he treated the other guys—that wasn't my business. And I didn't discuss this arrangement with any of my teammates. I just figured they must have known, basically, why I was gone.

I might have been wrong about that. Three years ago, I was in California with some of my old teammates, John Havlicek, K.C. Jones, and Sam Jones, for a charity golf tournament. Havlicek told me that during my second year, the team was practicing after the All-Star break when Red announced, "Russ is not here today." Just before they started the scrimmage, someone groused, "You know, we're tired of busting our humps scrimmaging while Russell's up there drinking tea in the stands or taking another vacation!" Red stared darts at the guy and said, "Okay. What team do you want to go play with? Because then you won't have to be here anymore to watch it." No one said another word. Then he announced to everyone, "Listen, listen, listen. There are two sets of rules on this team. One set of rules for Russell and another set of rules for the rest of you guys. Period!" And he walked away.

I had not known about that. I remembered asking Red one time, after we retired, if anyone ever complained about my disappearing for days at a time, and him saying, "Not really." Which wasn't a flat "No." This was something else I respected in Red, as a coach and a friend: his sensitivity to the feelings of each of his players. We were both careful not to say anything to each other that might compromise our outlooks on other players. We didn't want to do anything that might pollute how we all felt about each other as teammates.

I thought the germ of all this was planted when Red did all that hard work observing me and observing me, until he had the insight to ask me, "What's wrong?" When I told him, "I'm tired," he input that into his equation for me: "If he's tired, how can I counter that? I have to do everything I can to help him with that."

No one outside the team, not even the Boston beat writers, had a clue about how Red was coaching us, or what we were doing and why. I thought that was extraordinary in itself. It was also extraordinary how Red instituted such a versatile system for such diverse people and still consistently won. There was a general view that, since I was so pivotal to the team, and I had such a forceful personality, I was actually calling the shots. That was never the case. While Red never told me how to play, I never told Red how to coach.

In '56, Red had landed Heinsohn, K.C. Jones, and me in the same draft. Not a bad draft! Sometime afterward, I overheard Cousy tell Red, "There's no point in bringing K.C. to camp. I talked to some guys who played against him in the Army and they said he can't play." But Red always believed that he made good draft choices. Plus, K.C. was an amazing athlete. The Los Angeles Rams wanted him to play end for them, but he chose basketball. So Red brought K.C. to camp to take a look at him. And sure enough, K.C. made the team. Red's intuitive decision-making was fascinating and almost always on the money. It taught me to rely even more on my own intuition and instincts than I already did. I learned a lot more by doing exactly what he did to learn about me: I observed him at work as a leader.

Red didn't coach like anyone else and he didn't think like anyone else. His methods were thoughtful, purposeful, innovative, and fluid, and always geared to the team over any one player. Yet he also found ways to inspire trust and confidence in each of us separately. He could talk to a player two or three times and he would know how to talk to him from then on. And I never heard him say, "Boy, I'd sure like to have that guy on my team." I don't care how good the guy was. When other great players came into the league, like Jerry West, Wilt Chamberlain, or Elgin Baylor, he didn't feel he needed them. His attitude was: "Fuck 'em! This is my team right here." He never compared players. His focus was on the team he had, not the team he would have wanted to have: "These are the guys I'm going to war with. I'm going to win with what we got. How do I get the best out of *these* guys?"

That attitude reinforced how important Red considered each player, individually, to the team. He was telling us, "I have a system for the guys on this unit. No matter what your skills are, there's a place for you in this system. I will never say to you that you have to change what you did in college or someplace else in order to play for me. We brought you here because we think you've got skills. Now it's our job to find the right spot in the system for you." Most coaches back then operated the opposite way: "I don't know what you did when you played over there. But over here, you have to do this." In my mind, when you're dealing with men, that's a big mistake.

It might sound minor, but this approach told me something important about Red's sense of loyalty, commitment, and devotion. It was also an important secret to his success as a coach: he treated us like men working together as men. For one thing, he left ego issues to be worked out among the players. He didn't take those matters personally, or impose rules on us that said, "I'm the boss. You can't challenge my authority." Instead, he gave us all the personal responsibility that we could handle.

For example, he hated it when anyone came late to practice. To him, practice was part of the job description. His principle was *Give me a full day's work for a full day's pay*. Accordingly, he devised an impersonal fine system for the team that he never changed. He never let himself get personal because he didn't think the coach should affect the chemistry of the team. So he told us, "Here's how it works. If you come late to practice, don't say anything to me. No explanations. I don't need to hear it and I don't care. Look at the clock when you come in and pay the trainer a dollar for every minute you're late." It was up to us to govern ourselves on the job. Like men.

The reason he did it that way was that if a guy came late to practice and he had to make a specific issue of it, yelling, "Goddamn it! What's wrong with you?" that would be personal. This way, no one felt disrespected. The same approach also applied to how we practiced. Red never presumed to lecture us on our personal pride or sense of commitment. That was something inside you—he knew he couldn't impose it, even if he wanted to. He understood that when it comes to dealing with individuals, one size can't fit all.

Red's general attitude as a leader was that, as men, you either brought those tools with you or you didn't. If you didn't have a commitment to excellence, he'd see it and you'd be gone. He wasn't delicate. Sometimes, he even seemed a little cruel. But if you were on his team, if you weren't all about achieving excellence at your craft every second you were in the uniform, as he might say, "Flush it and leave." Even though he never explained his philosophy to us, it was always clear that it wasn't about what Red Auerbach or Walter Brown or the fans or anyone else expected from you on the Celtics. It was about what you expected of yourself. Which was exactly how *I* operated.

Red's philosophy also jibed perfectly with something about work that my father instilled in me as a kid. My dad told me, "Son, I don't know what you'll do when you grow up. But whatever you decide to do, I want you to be the best there is at that. What if you decide to be a ditch-digger? There's not much honor associated with that. But if you decide to be a ditch-digger, I want people in New York, Chicago, California to know, 'Hey, there's this guy in Louisiana digging ditches! You got to go to Louisiana to see *these* ditches!' What you want to do in your profession is go from a journeyman to an artist. And let everyone see your art."

An event at Boston's Fleet Center honoring Red. This moment occurred just after he told the crowd, "I wouldn't have been able to do it without the big guy over there."

Classic Red, working the sidelines, gathering information.

The 1957 Celtics, my first year and their first championship.

My way of saying, "Wilt, don't come in here with anything as weak as that."

The start of a typical Celtics practice session, just before I went to make a cup of tea and find my seat in the stands.

Another championship wrapped up.

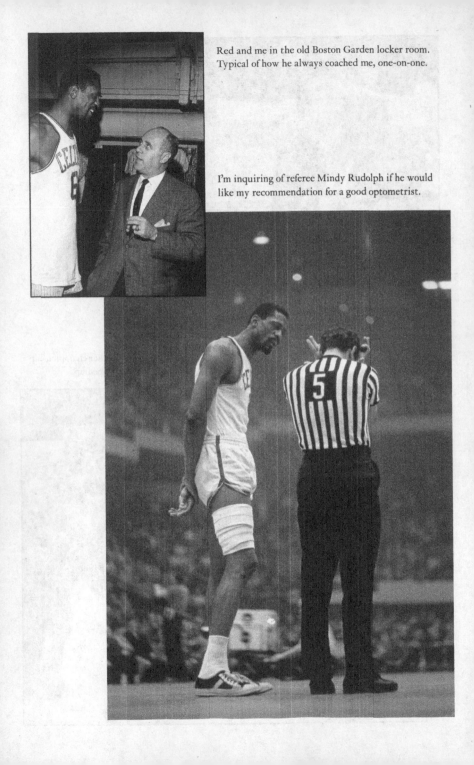

Red and me in the old Boston Garden locker room. Typical of how he always coached me, one-on-one.

I'm inquiring of referee Mindy Rudolph if he would like my recommendation for a good optometrist.

Just three of the many All-Stars who played for the Celtics (with Bob Cousy [L] and Tommy Heinsohn [R]).

The start of a Celtics fast break. We developed a team approach that was greater than any of its parts, from one end of the court to the other.

Another happy ending with John Havlicek. (See chapter 8.)

After I left, Red continued to win with the Celtics as general manager, in the '70s and '80s. Here, he hoists yet another championship trophy.

I spent much of my post-Celtics years working for charities, especially Mentor.

At Camp Millbrook, Red's summer camp, he checked out the rookies and his draft picks after the draft.

Red and me watching a Celtics game at the Fleet Center, years after our retirements. I'm poised just behind his right ear, like always, so I could continue teaching him about basketball.

Chapter 6

IIIIIIIIIIIIIIIIIIIIIIIIIIIIIIIIIIIIIIIIIIIII

We All Just Lived It

Philosophically speaking, men make friends completely differently than women do, even if the main elements—understanding, trust, respect—are the same. Most men I know don't talk about their feelings, or even, usually, about their friendship. They just let it unfold. The only time Red and I ever discussed our personal philosophies was when we talked about our work ethic, in the context of helping the Celtics win games.

I saw early how hard Red worked coaching us, and how nimble he had to be, mentally and psychologically, to keep us on track. He was an extraordinarily intuitive psychologist and motivator, and every now and then, he'd make a move on me. I always saw it coming, and I'd let him know it. But I knew it was always in the team's best interests, so I accepted it. Sometimes it even worked on me. For example, my second year I came out like a wild beast and we ate up every opponent, one after another. By the All-Star break, we had a twelve-game lead, and nobody had a prayer of catching us. Maybe that seeped into my psyche because, just after the All-Star game, Red called me into his office. I didn't know it, but I was about to get my first Red Auerbach pep talk. He lit a cigar and said, "I'm so mad, I could bite the head off a ten-penny nail."

"Red. What are you mad about?"

"We got the division sewed up already and we both know that."

I thought that was a *good* thing. "*That's* why you're mad?"

He said, "You're coasting! We got the big lead, so I can understand why you're letting up. But all you're doing is coasting just enough to get ready for the playoffs. Even *you* ain't that good. You can't turn it on and off in this league. You have to go hard all the time, Russ. Christ, you got these guys so terrorized, they can't play against you. But if you let up on them, and they start believing they can play against you, then they can *play* against you." He puffed at his cigar. "You know, at the end of the year, you should be the MVP in this league. But if you let up, and there's another player on the same page, he'll get it. So you have to take it off the table. Leave no doubt at all."

It was a masterful performance. Half scold, half flattering pep talk, in a soft, calm tone. He knew I was a very proud man. He was punching me to make me play with more pride, and tougher and meaner. He might as well have said, "Remember Kenny Sears!" He left something unsaid on the table, too. He was hinting subtly that the NBA was over 90 percent white, which we both knew was a factor in MVP selections. That night, I went out and broke the league record for rebounds in a game with thirty-nine.

This was how he coached me at his psychological best. He gave me just enough to think about and to motivate me at the same time. Everything he did was calculated. Instead of insisting, "This is how *I* want you to play," he was saying, "This'll be good for *you*" and, by association, "This'll be good for the team." We had a lot of these little conversations where he made sure he pushed the right buttons to help me. It was always something current—he never said, "Last year you did this." He never compared. We never talked about next year, either. We only talked about this year—what's going on right now. Even after we won our first NBA championship in 1957, all he would say about it was, "It's like yesterday's newspaper. The only thing it's good for is the bottom of a birdcage."

The amazing thing about how Red performed this constant motivational dance was how deft he was, and how deliberately impersonal, so that you would focus on the team, not yourself. I admired that about the man my whole life. It was part of our friendship too. Just as he conducted himself as a coach, as a friend he was always concerned about what was best for you, not him. How many friends like that do you have in a lifetime?

Red worked me over this way all season, sparking me here and there, trying to keep me ferocious to win, just like he was. As always, his real target was implied rather than overt. One time, he casually remarked that an opposing player had been on a run against me lately, playing great. I barely heard it, yet it marinated until I wondered, "Now, when did that guy *ever* play me good?" Bang! I couldn't wait to get to that guy the next time we played. *But if you let up on them, and they start believing they can play against you, then they can* play *against you.*

In one game in the mid-1960s, Bill Bradley, the gifted small forward on the New York Knicks, was having a particularly good night against my teammate Satch Sanders. Satch was a very good player, but on this night, Bradley kept finding ways around him to make all these open shots. Now, I was on the floor trying to help my teammates. But, just like Red on the bench, I couldn't help Satch physically, so I decided to try my own psychology. I always strived to play what I called The Perfect Game. My perception of The Perfect Game involved a whole bunch of criteria: shooting percentage, free throw percentage, total rebounds, total blocked shots, assists, screens—and conversations. Why conversations? The power of language.

I took Satch aside and said, "You know, Satch, on the uniforms there's a big number and a small number. The big number's on the back. The small number's on the front. I know you haven't seen that small one yet, but trust me, it's there." Translation: "That's Bradley's back you're seeing as he goes by you. Front him more, so he can't get past."

It didn't help. In the last quarter, one of the Knicks was at the free throw line, ready to shoot a foul shot. I was standing on one side of the foul lane and Satch was on the other side, right next to his friend, Mr. Bradley. I saw these two guys standing together and I thought, "Something's wrong with this picture." I was team captain then, so when the guy prepared to shoot the foul, I called to the referee, "Hold it up." The referee took the ball back, and I stepped slowly across the lane to Satch and looked him in the eyes, hard. Then I talked just loud enough for him and Bradley to hear. "Satch, can you guard this motherfucker?" I rarely used that word. I was telling Satch, without saying it, "This guy you're guarding isn't just another ballplayer. He's a *motherfucker*! He has no respect for you! He thinks you can't guard him!"

Satch grunted, "Yeah, I can guard him."

I said, "Well, goddamn it, *do* it!" I turned around and stepped back to my side of the lane. That performance was for Bill Bradley's benefit—it had nothing to do with Satch. In fact, Satch didn't do anything differently after that. But Bradley did: he had a lousy final quarter, and that was a big reason we won the game. Years later, when Bill Bradley ran for president, I went around the country with him. In Iowa one night, he reminded me about that incident—it still bothered him! He said, "Russ, what the hell *was* that?" I told him it was designed to get inside his head. He said, "It did. It threw off my concentration and I couldn't do anything right the rest of the game." I said, "Courtesy William F. Russell, Doctor of Psychological Warfare!" And we shared a good laugh.

What Red accomplished from the bench—and this was an overlooked key to his coaching success—was an important part of why we respected and trusted him so much: he was constantly working to help us. There was not only an element of genius in his efforts on our behalf, but also of his integrity as a man. It was like my father's admonition to give three dollars worth of work for two dollars pay—Red was giving us *ten*! And we recognized that it wasn't for him; it was all for his team. In his selfless devotion and loyalty to us, there was a measure of almost tender affection that men don't often share in their relationships with other men. Certainly not in our era. We sensed it, although we never mentioned it. We just lived it. That's what men did.

Red was more animated on the sideline than any coach I ever saw. In fact, his screaming fits and emotional tempests are what most people think of when they think of Red Auerbach. He was never content to sit on the sideline and watch us work. He was always calculating how he could help. He knew he couldn't be out there on the floor with us, so he was on his feet all the time, pacing with a rolled-up program, pounding his free hand with it as if it were the other team. He kept busy as hell, yelling at the officials, "Damn it!" and "Bullshit!" and whatever else he could think of to mean, "You lousy bastards are screwing my team!"

For forty-eight minutes, Red stalked the sideline back and forth, busting the referees' chops, cursing the other team, bitching at the scorer's table, challenging the timekeeper. He was just as intensely involved as we were, trying to leverage any little advantage he could. For example, if we weren't going so good, he would find ways to alter the pace of the game. His favorite last-resort ploy was to deliberately get himself ejected. He was an absolute master at that, right up to his last year as a coach when he got himself tossed from sixteen games, including the All-Star game. It not only took the other team out of their rhythm and flow, but it also gave us a chance to regroup and refocus.

I remember, especially my rookie year, that whenever I looked over there, I saw more than just Red Auerbach raising holy hell. I saw a community. To me, in all his passionate devotion and striving, he represented not his tribe, and not my tribe, but *our* tribe: *the Celtics*.

I always said that the foundation of Red's coaching method was mathematics: *Do anything you can to change the odds in your favor.* He would do absolutely anything to accomplish that. I don't think the officials knew this, but he used to scout them. It was fascinating because Red, philosophically, refused to scout even his opponents. I asked him one time, "How come you don't scout?" His reasoning—which I loved—was "Fuck 'em! I don't care what *they're* doing. Let them worry about what *we're* doing!" That was his bulldog take on life: you figure out what you want to do and how you want to live, and you go do it your own way. That was another reason we became friends—I was already leading my life this way when we met.

Red arrived at the arena early to find out who was officiating our game. Then he'd tell us in the locker room, "We got Lou and Stan tonight. Lou's a pussy. He calls traveling a lot. Stan's a prick. He won't call elbows. But he'll call a foul if you fart." Red input the tendencies of every official in his game-day equations because he felt it might give us an edge. During games, he kept trying to push that edge by letting the refs know he was eyeballing everything they did, from the way they ran around watching the play to how slow or quick they were to blow a whistle. He knew their habits and quirks, so he could anticipate what they'd do in almost every circumstance, even before they did it. It was just another way for him to actively help his team. It was like when he protested that goaltending call against me: *If every time they make a tough call against us, I raise holy hell, they might think twice and change their call.*

He didn't just razz the officials. He tried to intimidate them psychologically. For years, Red had a running problem with a ref named Sid Borgia, who made a lot of terrible calls against us. When Sid had our game, it was almost as much fun watching the fireworks between him and Red as it was winning the game. Red always smoked these foul-smelling cigars—in those days, people still smoked indoors. So, one time when Sid made a lousy call on us, Red formed a mouthful of chewed-up cigar and got right in Sid's face and started arguing, "You son of a bitch!" and "Jesus Christ!" and Sid ended up with a face-full of chewed tobacco bits. I don't know if this was deliberate. But, after that, the moment Sid saw Red heading his way to argue a call, he retreated farther onto the court, out of Red's range. Then he and Red traded the vilest curses I ever heard. I don't know if it made Sid think twice about making calls against us. I do know that he hated doing Celtics games while Red was our coach.

Red used to say, "I don't give a damn if it works. As long as I can distract them from their tendencies." Along with the big artillery lobs at the refs, Red used small-arms fire. This included yelling "cuts" loud enough for specific opponents to hear: "Give Russell the ball! This *asshole* can't guard him!" "Let's get Sam a shot! He ain't got *nobody* on him!" "Don't sweat this guy. He can't hit a bull in the ass with a banjo!" It was adolescent-funny at times, like playground games where you said whatever it took to get under the enemy's skin. It wasn't about what you said to him. It was about the tiny chance that it might distract him momentarily from the task at hand.

The point was, to Red, almost nothing was too insignificant to consider when it came to seeking an advantage. Fred Schaus, who coached the Los Angeles Lakers from 1960 to 1967, despised Auerbach. Red got so deep under his skin that, at the end of a game with us in the lead, when the Lakers huddled to talk strategy, Schaus used to sneak glances at Red to see if he was lighting up his famous "victory" cigar. Schaus was offended by that gesture because, when Red lit up, it looked as though he thought the game was over. Schaus considered it arrogant. He told Red one time, "I want to come over there and make you choke on that cigar!" That was all Red needed to juice him up even more. He figured if Schaus was giving that much thought to him and his cigar, he wasn't busy coaching. So he kept feinting with the cigar at the end of our games with Schaus, just to distract him from coaching his team and to irritate him for the next time.

I started picking up on that, too, because I wanted to win as badly as Red. One night, I had a hamstring strain, and the trainer didn't get around to putting the wrap on it, so I went out for warm-ups without one. When they introduced the lineups, I walked onto the floor very slowly, instead of running out like usual, because I didn't want to pull the muscle. I noticed that the other team looked pissed, as if they were thinking, "Who the hell's he think he is? How come he's not running out like everyone else?" When I realized how much this bothered those guys, I never ran out during introductions again.

I added other touches, such as walking out there slowly and scowling and folding my arms, as if to say, "I'm here." That was part theater, of course, like Red and his cigar. Pretty soon, Red and I were doing these things in tandem, like a couple of tag-team wrestlers. We used to talk about it after games, which sometimes was enlightening. When I told Red, "You know, it's really bullshit when I walk out for introductions. I didn't do it at first to piss anyone off," he said, "Yeah, I know what you mean. The first time I lit a cigar, I was just lighting a goddamn cigar! When it bothered the shit out of them, I made it a part of my personality."

I realized very early that I was fortunate to be privy to his work, and to understand and appreciate what he was trying to do. And I knew that he was watching me work and appreciating what I was trying to do. There's a famous show-tune line about a "mutual admiration society." That's what Red and I had. Once he started talking to me about his favorite strategies and targets, the friendship started to bloom.

One of Red's favorite targets was someone you couldn't possibly miss: Wilt Chamberlain. Wilt was a great player and competitor, and we eventually became very close friends. But, like Red and me, he wore his pride up front, and he took no crap from anyone. Mix that with Red's Brooklyn, tough-guy attitude and you had psychological TNT. One game, Red started jawing at Wilt, who was playing me tough. He ragged on Wilt, yelled at him, cursed him up one side and down the other—anything to rattle him or throw him off his game.

To me, it was like hitting an elephant in the ass with a pebble. Yet, somehow, Red succeeded. All of a sudden, there was Wilt looming over Red on the sideline, a couple of feet away. Wilt was, by far, the strongest man in the NBA—even I didn't want a piece of him. But when he looked like he might take a potshot at Red, I hurried over and stood beside my coach. Red just kept on hurling pebbles.

I said, "Red." No response. "Red! You're too close."

He said, "Leave me alone. I ain't afraid of that big son of a bitch. I'll kick his ass!"

Red had told me once, "Brooklyn made me so tough, nobody ever kicked my ass." I believed him, because he would fight at the drop of a hat—he was always throwing sombreros around, really *big* hats. Plus, I had never seen him bluff or back down from a confrontation. He was a genuine tough guy. For example, there was an incident in St. Louis my rookie year, when we were playing in the NBA finals. Bill Sharman was very exact about everything, so before the game he measured the baskets and told Red, "That basket ain't ten feet high." So Red got the workers to measure. As they were measuring, Ben Kerner, the St. Louis owner Red had always disliked, charged onto the court yelling at Red for trying to make him look like a cheater. Red hauled off and knocked Kerner's two front teeth to the middle of the court—and never said a word. I thought of that incident as Wilt grew angrier with Red.

I tried again. "Red! Goddamn it!"

"*What?*"

"You're standing *too close*! He can swat you from there." Red hadn't noticed that while he was busy getting under Wilt's skin, he was about to be under Wilt's fist. "Take a couple steps back, and *then* have at it."

He stepped back slowly, and I thought it was over. But then he unloaded insults again on Wilt. The referee came over and said, "Okay, break it up. C'mon, you guys, let's finish the game." That defused things, and we all turned around and returned to the game.

When I hurried over to Red's side, it was the first time I realized that I wouldn't let anything happen to him. I was protecting him *personally*. It struck me that being a friend was now part of the deal—I was making sure that *my friend* was out of harm's way. In hindsight, my warning him, "Take a couple steps back," was like his warning me, the last time I saw him alive, "Don't fall."

After that, I was always aware of protecting him. It came up often, and I made no pretense about how I felt. Two particular examples come to mind. The first happened in a game against the Knicks. Red was at the scorer's table, bitching and moaning about something. Whenever he did that stuff, I sat down on the bench and waited until he ran out of gas and we were ready to go back out and win the game. But then Harry Gallatin, a former player coaching the Knicks, walked over to Red and barked, "Red! Let's finish this goddamn game! Go sit your ass down!" I got there immediately and called out to Harry, "Who the hell are you talking to? Red? The *little* guy? *I'm* your size—talk to *me* like that, goddamn it! You know you been a coward your whole goddamn life!" Harry just turned around and walked back to his bench, like he was supposed to.

The second event was a fight that broke out at Convention Hall in Philadelphia. While both teams were fighting, fans climbed out of the stands, and one of them got behind Red and pushed him hard in the back. I came up behind this guy, grabbed him by the neck and head, and hauled him around, with his feet off the ground, trying to stop the fight. Chet Walker, a Philly star, saw this and yelled, "Hey! Russell's *killing* that guy!" After everything cooled off and I let the guy down, Red came over to me, all juiced up, and said, "Thanks, Russ, goddamn it! That son of a bitch—we can't let him get away with that shit!" He looked like he wanted some more!

Now, here's a little subtlety about how we handled this aspect of our friendship, as it grew. After the Wilt thing, Red came over to me and said, "Thanks for looking out for me, Russ." I didn't want him to feel beholden to me. No more than he had wanted me to feel that way when he came to my defense on the goaltending incident. So I said, "Bullshit, Red. I wasn't looking out for *you*. I didn't want *him* to get arrested for manslaughter!"

That was part of our unspoken code: we were working together now as men, not just as co-workers. That's what friends do for each other: they always have each other's back. No need to carry it farther than that: *That's the way it's supposed to be done.* But from The Wilt Moment on, I always felt that Red knew in his heart that if Bill Russell was in the building, he would never let anybody do anything to harm him.

||||||||||||||||||||||||||||||||||||||||||||||||

Chapter 7

Boy, What Fun

Ο ne thing I learned from the Jesuit teaching methods of my college was that the professors trusted their students with knowledge. There was nothing they were afraid to let us learn. I had a class—I think it was Poli Sci 140—that was a study of communism. We were required to read about Marx and Engels and that bunch. We were encouraged to go hear the longshoremen of San Francisco speak, men dismissed by many as "Commies" or "Commie sympathizers." The school even arranged to bring some ex-POWs from the Korean War to talk to us about how the Communists had brainwashed them.

In essence, the Jesuit theory was: *We want you to learn everything we can expose you to about communism. Once you recognize it for what it is, we don't have to worry about you becoming a communist.* Pretty wise, I thought, because they created an atmosphere of learning so you could improve yourself.

Red's coaching reminded me of that approach, which I think is also a cornerstone of a successful friendship. What you don't know—about your teammates, your coach, your friend—can hurt you. For a coach, each player has his own little red wagon. How do you satisfy all these individual agendas? Red's solution was deceptively simple, yet remarkably rare: instead of fighting them, or trying to change them, he embraced them. In other words, he made his system versatile and fluid enough to incorporate *everyone's* little red wagons. He understood, as I did, that the more platforms you provide for people to succeed with the skills they already have, the more motivated they are and the more likely they *will* succeed.

Once we accepted that we were a unit, Red expected us to behave like one, on our own. But he understood that, as coach, he could not be the one to enforce that issue. He would set the table; the rest was up to us. This was not just basketball wisdom. It was a lesson in leadership, in respect for individuality, and ultimately, in friendship. In effect, Red created the environment for open collaboration. It was like what film actor Oliver Hardy used to say to his sidekick Stanley Laurel, "Why don't you do something to *help* me!" That was Red's approach: *Here's something to help you. Now go make it work.*

"Russ," he'd say, "I think we have to get Sam or Heinsohn to do something. How do *you* think we should get it done?"

Then we'd trade opinions. Sometimes, he'd say, "Russ. I think this is the time in the game, if we can shut them down for the next three minutes, we'll have the game won. That means they only get one shot—and not a good shot. No open shots and no offensive rebounds." Sometimes, I'd accept that he knew what he was talking about and I'd go out and make sure the team put that strategy in play. But if I didn't agree with a particular approach, I had the option of saying, "Well, I think we should do it this way," and he would almost always say, "Okay. We'll do it your way." Not begrudging, simply obliging—because he'd learned, in situations like this, that I knew what I was talking about, too.

Red found subtle ways to transfer to us the responsibility for things he didn't control that he knew we had to do to win. He never imposed anything. Instead, he set it up so that we had to buy in to the system and then be responsible for whether the plays worked or not. I was always conscious of the fact that I was largely responsible for any game's outcome. If I played well, and I dominated on both ends, we won. If I didn't, we almost always lost. I felt the weight of that responsibility every game. What he said was, essentially, "This is the result I want for the team. You know how to get it done for us. Go and get it done." And that's what I would do.

Here's an example of how I went about it. Bob Cousy was a great teammate. But in our approaches to the game, he and I were complete opposites. I approached the game defense first, then offense. He approached the game offense first and then, maybe, some defense. I never had a negative thought about that—he was always a great team player and a really good guy. But you can't win championships with just an offensive approach. Over our time together, I think I helped him appreciate that because, years later, he told me that his most lasting memory of our first championship together was a defensive play I made. After I watched him play a dozen games or so, and had a few brief conversations with him to learn more about his approach, I figured out how to help him to help us win more games. If we were on defense and his man got by him, I'd slide over and pick him up to keep him from getting a layup. On offense, after I'd get a rebound, I'd start deliberately looking for Cousy to get him outlet passes to start fast breaks. Probably a third of the shots I blocked for Cousy were off his men. But I never criticized him, behind his back or to his face.

The way I looked at it was that his flaws on defense triggered our offense. I'd block a shot and outlet it and get it going the other way, turning it into a strength. Also, it demonstrated to the other guys that if their man got by them, I'd be there to back them up, too, and I would never complain about it. Instead, I would find a way to incorporate their defensive lapses into our offensive system. I think that was one of my most gratifying contributions to the team. And Red perceived that.

After I'd been there about a month, Cousy walked over to me at practice and said, "I know when you get a rebound, you look for me. So what I'll do is, after they take a shot and you rebound it, I'll go to that spot over there. Look for me there first." He had started to appreciate what I was doing for him. By the end of that year, our team scoring average was the highest in franchise history. That was based on my defensive rebounding and outlet passing, mostly to Cousy. The other team would shoot and, four seconds later, we'd be making a layup at the other end. But, really, all I did was help Cousy do better what he did best.

It's the same dynamic that creates friendships. If you ask yourself, "What can I do to help my friend?" you will automatically do it, and be a good friend. Here's the quintessential Red Auerbach example of how to be a man among men, and how to be a friend. My early years with the team, we had a player named Frank Ramsey, who was a dead-eye shooter from the University of Kentucky. But in the pros, he was labeled a "Tweener"—too tall to be a guard, too small to be a forward. In those days, players were classified as either "First Team" or "Second Team" material. If you were tagged "Second Team," it meant that you weren't good enough to start. The fact was, Frank Ramsey was a great player who was all business and who could have started for a lot of other teams. But when he came to the Celtics, he had to come off the bench because we had five solid starters ahead of him, and Red was loyal to his starters. With Red, you worked your way to starter status and, once you got there, the job was yours until you stopped producing, got hurt, or retired. It was an unwritten code with us, so that no one competed actively for anyone else's job. If you played for Red, you waited your turn.

If Red had said what almost every other coach would have said back then—"Ramsey doesn't fit the prototype for either position"—Frank never would have had the opportunity to become a great player. But Red didn't say that, because he was blind to the conventional wisdom. He also understood that if you stereotyped people, you'd never see past that. All Red cared about was what you *could* accomplish, not what someone else thought you *couldn't*.

When Red put Frank Ramsey in a game, one thing was certain: Frank would shoot. No matter where he was or what the circumstance, the second he touched the ball, it was going up. He didn't do that because he was selfish or anxious or dumb. He didn't have an attitude about it: "This is what I'm going to do." No attitude at all. He *had* to shoot; that was just who Frank was. It could have been a big negative— guys can get pissed off when the guy off the bench bypasses all the starters and shoots. Suppose Cousy gets Heinsohn three shots, and he hits all three in a row? Now Heinsohn's on a roll. Then Red puts Ramsey in for Cousy at guard, and Frank gets the ball and, bang, he shoots. Heinsohn says,

"Jesus Christ, Frank! I was going good! Why don't you go up and down the floor a few times and get a little loose!" On another team, that would disrupt the offense. Not on Red's team. To Red it made sense. That was why he never complained about Frank shooting willy-nilly. He just watched how Frank liked to play, and listened to him, and then one day he told him, "You're my Sixth Man." In other words: "You're not First Team. You're not Second Team. You're no sub. You're my Sixth Starter."

It used to be that when a player came in off the bench, you worried he might be a liability: "Let's hope he's not a step backward." But when Frank came in for us, he was an asset—a step forward, psychologically and physically. I thought the genius part of this equation was that instead of fighting Ramsey's tendency to shoot, Red embraced it and developed a whole culture around it. Pretty soon, when Frank got up to come into a game, there was a sense of anticipation that had never existed before: "Here comes the big shooter!" So, now Frank Ramsey, former "Tweener," was respected and even feared around the league as the Celtics' dangerous "Sixth Man"!

Red not only made Frank comfortable coming off the bench, he made being *first* off the bench an honor. Psychologically, this got the whole team involved in "honoring" Frank. Our attitude became, "Frank's going to shoot, we know that. So let's try to get him some good shots." I made it my personal responsibility to make sure he *always* got a good shot by passing him the ball where and how he needed it.

It helped us in the locker room, too. Before every game, everyone started betting a quarter on how long it would take Frank to get off a shot after he stepped on the floor. We loved that he was such a feared shooter. That was all Red's doing. His ingenious Sixth Man concept made Frank Ramsey a full-fledged member of our team and more famous than some of the starters, all while improving our chances of winning. Not only that, but Red had obliterated the "First Team–Second Team" prototype, knocking it, he would have said, "ass over tea kettle!" It's fifty years later now, and have you noticed that the NBA presents an annual Sixth Man Award?

What Red had done for Frank is what a friend would do for a friend. I always gave credit to Red for creating this open, free-flowing atmosphere. It transcended basketball and improving your skills as a player. It was also about life, and learning about each other as men, so that, ultimately, you could become a better teammate and a better man at the same time. Where do you get that at work these days? We had it year after year. No one else outside the team perceived this, but in Red's system, once we all started learning about each other, teamwork became a tangible expression of friendship: "I will back you up and you will back me up. We will watch out for each other and help each other, and that is how we'll win."

Some people have asked me, "What's the difference between Phil Jackson winning nine championships and Red Auerbach winning nine champhionships?" Trying to put my bias aside—and there's a great deal of it—I think Red's achievement is superior because he got the same team to listen to him for ten straight years, and not just win a lot of games, but, at one stage, eight consecutive championships. What does a coach say to a player, ten years into their relationship, that he hasn't heard a hundred times? There's a cliché, "Familiarity breeds contempt." Well, with Red and us, it worked just the opposite. To me, that took more than just a great coach. It took a mastermind. It took, in fact, exactly what you also want in a true friend: someone big enough to consider his friend's needs first and find the right ways to help enhance his existence.

Our synchronicity on Red Auerbach's Celtics was exhilarating. And so unique. And, boy, what fun.

My whole life, I have always been the most private person I know. I drew that line when I was very young. I considered uninvited intrusions into my privacy extremely disrespectful, and I would not tolerate them, even from a coach. Fortunately, Red brought a flexible sensibility with him from his tribe, and he never tried to regulate our private lives. Well, almost never. He instituted only one curfew as our coach, and he immediately regretted doing it.

Mid-season one year, we had won seventeen straight games, and the eighteenth game would have broken the NBA record for consecutive wins. We were playing the first game of a doubleheader against the worst team in the league. Red made it clear it was important to him that we win this game. So the night before, he called a curfew: everybody in their room before 11 P.M. He said he would call our rooms and we had to answer the phone to verify we were there. We did that, and everybody got a good night's sleep. Next night, we went out to play the game and we got the hell kicked out of us.

Red came into the locker room afterward and said, "I blew it." He realized that everyone had developed his own routine to get ready for a game, on his own internal clock, and that the curfew wrecked all that. If you're doing your work, I cannot, as an observer, tell you how many hours you should put in or what time you should get started. Sometimes, you don't lose a game so much as you just get beat. Red figured *he* got us beat. And he was man enough to admit his mistake in front of his team. It was the first time I ever saw a coach apologize. He didn't let ego or false pride trump personal responsibility. I admired his willingness to share in the blame: *Win as a team, lose as team.*

It was interesting also because that was part of the Celtics culture under owner Walter Brown. Everyone in the organization loved Walter. It was not a surface thing, it was genuine—he was the heart and soul of our 1950s and early 1960s teams. His attitude and character were the foundation for a coach like Red, who followed his own instincts anyway. It became obvious that Red was under no pressure to please Walter, because Walter always backed him up. That turned out to be very big for me because I could not have thrived in any other atmosphere. With good men like Walter and Red running the club, the Celtics became an island of refuge in a turbulent sea.

The tone was set long before I arrived. In 1950, Red decided to draft Chuck Cooper, who would then become the first black player ever drafted into the NBA. Red gave that absolutely no weight. He wanted Chuck Cooper for one reason: to improve the team. Walter agreed without a second thought. But, at a meeting before the draft, a couple of owners tried to persuade Walter not to go through with that pick. One actually said, "He's *colored*, you know." Walter said, "I don't care if he's striped, plaid, or polka-dot. That's the player we want." When Chuck Cooper came to Boston to sign his contract, Walter shook his hand and said, "Mr. Cooper, I want you to know the Boston Celtics will never embarrass you."

So now, ironically, in a city that was rife with racism, the Celtics had just become the most progressive franchise in basketball! That was mainly because Walter and Red were decent human beings who treated men as men, not pawns or pennies you put in a slot. They had very different temperaments: Walter was turn-of-the-century polite, pliable, and gentlemanly; Red was blustery, blunt, and tough as oak. But they were two sides of the same coin when it came to putting the best players available on their team—white, black, or polka-dot. Their integrity, class, and commitment never wavered while they ran the show together.

Walter was also obsessed with loyalty, something my tribe valued enormously. At the end of my rookie year, we lost a pivotal regular-season game to Syracuse. Two days later, we were scheduled to play them again in the playoffs, where they had dominated Boston teams in the past. We dragged into our clubhouse and there was Walter Brown in the middle of the room, red-faced and puffed with rage. "You bunch of chokers!" he uncorked. "I pay all that money for you to choke like this? These guys are going to beat us *again*! I'm so mad, I will never come back into this locker room with you overpaid chokers!" His tirade astonished us because everyone knew Walter as an easy, gracious soul.

Before the playoff game with Syracuse, we were in the locker room getting ready when in walked Walter. He took off his fedora, looked around sheepishly, and announced, "Men, I want to apologize. I'm just such a huge fan that I got frustrated because these guys always eliminate us. But I should have never said it to you guys. You didn't deserve that at all. You're a terrific bunch of men and a great team. I apologize to all of you very sincerely, and you have my word I will never do it again." He put on his hat and walked outside.

We had already shrugged off his big blast—we all knew who he really was. But his apology was significant. This was our owner—in fact, the founder of the NBA—and he'd blown his top over a tough loss, then behaved with honor and dignity, and apologized to us. I thought, "This is a man you can trust and respect. He's secure enough in his own skin to take responsibility for his own mistake." That's another key ingredient in a true friendship: good friends must be confident in their own skin. One reason friendships sour is that, while they're developing, some people do things out of character, trying too hard to put their best foot forward. They aren't confident enough in who they are to know that should be enough. So they try to become someone "better." Both Walter and Red personified being who you are.

With Walter behind him, rubber-stamping his requests and trusting his judgment, Red was free to run the team his way. He was his own man, doing what he was compelled to do, instead of what somebody told him to do. In effect, Walter freed Red to be my friend, which enabled Red to accept me as I was and how I played.

Another trait of Red's that made him so successful, not just as a coach but also as a man, was that he knew how to communicate so that you always heard him. In the NBA, even today, when a coach fails to communicate with his players, it's usually because they've tuned him out and can't hear him anymore. Red never had that problem with us. It took real brilliance to keep his teams winning consistently and still keep the players listening. If we played really crappy—which we did some nights—Red would come into the locker room, grab his coat, and leave without a word. I asked him once, "Why don't you say anything after we stink up the joint?" He said, "After a loss, if it's a bad loss, nothing you say will penetrate. They can't hear you. So, why bother? It'll only get us all upset and that's not constructive."

To Red, "constructive" meant focusing only on winning. If we were losing badly in the fourth quarter, he was already thinking about the next game: "Okay, we lost this one. That won't change. Now, how do I keep this loss from affecting the next game?" He turned all his thoughts to getting us ready to win that one. When he did speak to us after a loss, or even just a lousy execution on a play, he knew that if he said the wrong thing, somebody might tune him out and never hear him again. He also knew that, for some players, their self-esteem was determined by how well they played. If they didn't play well, they thought, "I'm a bad person." "I have no guts." "I'm not trying hard enough." If they had swallowed that stuff since childhood, hearing it from coaches could only reinforce their insecurities. Red tried to motivate such players so that, next game, they would try to be "a better person." Red's approach was simple, really: he talked to everyone differently because he recognized, and respected, that everyone *was* different.

Another of Red's master strokes was knowing how to treat each player differently and still retain our cohesiveness as a team. For example, Red never yelled at me. He knew that Bill Russell would never respond to that. He yelled at Satch Sanders and Tommy Heinsohn because they responded to being yelled out—they needed it sometimes to get motivated. But that wasn't intimidation; it was just yelling. They understood that, because Red made every player secure in his role. "These are my starters. They earned their spots." *These are the guys I'm going to war with.* So the guys he yelled at never thought, "He'll trade me now because he's yelling at me." On occasion, though, he might try some sleight of hand, if he thought it would help the team. For example, he knew that since everyone knew that he never yelled at me, this in itself sent a message. My second year, I won the league's MVP award. The night before our first scheduled practice for the following season, we were at a team dinner. Red took me aside and said, "Listen, Russ. Tomorrow morning, nine o'clock, we'll get started. I want to make a slight change this year. I want to yell at you a little bit. I want to scream bloody murder, insult you any way I can. Be up your ass. I'll have to yell at some of these guys, but if I can't yell at you once in a while, I can't yell at them or they'll feel persecuted. But listen, when I yell at you, don't pay any attention to it. It's not for you, it's for them. And do me a favor, don't get pissed and go off on me. I just wanted to ask you first if this is okay with you."

It was okay with me—I thought it was intriguing, and I appreciated his sharing the plan. Next morning at practice, he ripped me up one side and down the other: "Russell, you son of a bitch!" this, and "Goddamn it, Russell!" that—every vile curse in the book, and a bunch more from the Brooklyn sewers. I was so blistered, I almost broke the ruse. But I controlled the urge and, later, I had a chat with Red. I said, "You heard about guys that are given an unlimited budget and they go *over* it? Well, you damn-near used up *all* my goodwill! You came that close to me coming over and slapping the crap out of you!"

We both laughed. He said, "If you had done that, Russ, I would've had to kick you off this team! *Then* where would we be?"

"I know where *I'd* be, Red," I teased—a little friendship thing going already. "*You'd* be up Shit Creek, without your goddamn paddle!"

Chapter 8

||||||||||||||||||||||||||||||||||||||

Godspeed

One day in 1966, Red called me into his office and told me, "This is it, Russ. This is my last year. I'm retiring."

My first reaction was, "Why don't you give it one more year?" I asked only because I wanted him to know how I felt. I was also being selfish—I wasn't finished with basketball yet and I wanted him to coach us to another title.

"Nope. I've had enough."

"Okay," I said. "Godspeed."

Out of respect, I didn't raise the issue again. Ever. Part of our friendship was that we never tried to coerce each other into doing anything the other person didn't want to do. It was never, "What you want to do is less important than what I want you to do." It was the other way around.

Red sat back and unwrapped one of his cigars. "*You* want the job, Russ?"

"Hell, no! After what I watched you go through? Screwing with the referees, the timekeepers, writers, fans, all that stress? I don't want any part of that!"

"Well, I have to hire a coach." He lit up his cigar and started calculating a new equation. "But I'll tell you here and now," he said, as though he missed me already, "I will not hire anybody to coach unless you give me your hundred-percent approval. We've been through too many wars together. You played too hard and meant too much to this franchise for me to leave you behind to deal with somebody you don't like. I don't want anybody messing with you."

This wasn't my coach talking anymore. This was my friend. And my friend would not leave me out there, subject to the whims of a coach who didn't know me. It sounded like the Marines: Never leave behind your brothers-in-arms in a combat zone. How many coaches or general managers thought that way about their players back then?

I said, "I appreciate that, Red." And I did. But sometimes, you have to think about yourself, and only yourself. I was thinking, "Well, this is breaking up that old gang of mine. How will this impact me? Will I be able to get along with the new coach?" I thought about the fact that everyone around the league knew that Bill Russell didn't always practice. A lot of Red's fellow coaches felt strongly, "If Russell played for me, he'd be practicing like everyone else!" None of them knew anything about the way Red and I had worked together—how he always tried to help me, and how we ran our plays.

I finally said, "I don't know what I'll do without you. I can't think of any other coach that I would want to play for."

"Okay," he said, "here's the deal. I want you to go home and come up with a list of five names. Guys you respect. I'll come up with five names, too. If someone's on both lists, we'll talk about him."

Here was probably the greatest coach ever, answerable to no one but himself, collaborating with a player to choose the team's next coach. It was an enormous compliment. I was losing more than just a great coach. I was losing the person who knew me, as a player and a man, in ways no other coach could fathom. I knew I could play for a coach I did not respect—I proved that in college and the Olympics. But it would be much more difficult.

I once remarked publicly, "Whenever I leave the Celtics locker room, even Heaven wouldn't be good enough because anyplace else is a step down!" I couldn't imagine anyplace more pleasurable to work. With Red and Walter Brown, I was the freest athlete on the planet. I could always be myself with them and they were always there for me. And Red and I could always work out our differences without friction because we let each other *know* about our differences. It wouldn't be easy to find someone to fill his shoes.

We both went home and drew up our lists of coaches. But we had two different agendas: mine was to find a good man that I knew could coach this team; Red's was to find someone who wouldn't mess with Russell. When we met and compared the lists, there were no matches. We still didn't have a coach. I said, "Okay, I'm going home." Next day, Red called me and sounded relieved. "I got a name. This is the guy I want to hire." He told me the name. "What about him?"

I didn't have to think about it. "No!" I said. "I can't play for that guy. If you think you'll hire him, I'm retiring with you. I don't even want to be in the same room with that son of a bitch." The guy he named was a successful, veteran NBA coach. But I had read things about him that I didn't trust. He'd also coached a good friend of mine, one of the greatest players in the game, who had told me how this guy tried to manipulate him in disrespectful ways, including trying to persuade him not to talk to me during the season.

I had even more disturbing facts. I've been very fortunate in my life because I've always been able to gather a lot of information from a vast array of sources. Since I had made it a point to communicate with black players on every team, I knew how every coach treated his players. Another of this guy's players had called me that same year to ask for advice. He said, "I got a problem. Coach just told us, 'I may be old-fashioned, and I may not be right, but I can't stand it when black guys date white women.' The problem is, I'm married to a white woman. What happens when I show up with my wife?"

I said, "Here's what you do. In training camp and exhibition season, you play as hard as you can, and do as much as you can to help win those games. Okay? And then, if he says anything to you about your wife being white, you tell him to go fuck himself!"

Red knew about none of this; he had no idea what kind of man this guy really was. But after I said I didn't want to be in the same room with him, Red didn't need to hear anything more. He just said, "So, what do you want me to do?"

"I don't know. I'll call you tomorrow."

This guy had been Red's first choice. Yet he didn't challenge my rejection. In a true friendship, nothing about your principles is up for debate. If I'm your friend, why would I *want* to debate you on your principles? All that mattered to Red was that I had objected: *If either of us asked a question, and there was a yes-or-no answer, both answers were acceptable.* It may seem like common sense. But, throughout my career, a lot of people had difficulty appreciating the wisdom of this commonsense approach.

My father had never explained the key decisions he made in his life. His principles were clear, and we accepted them. Often, today, people are encouraged to state their principles, to say, "This is what I believe." But for me, friendship is simpler than that. *You don't talk about it. You just live it.* A true friend accepts you, no questions asked. It's part of who you are, and that should be good enough. Otherwise, you're like those married couples who keep trying to make each other over. Mutual, unconditional respect is rare in relationships these days, never mind between naturally contentious men. That's one reason I have a permanently limited number of real, true friends.

That night, I thought it through and concluded that I had Red over a barrel and shouldn't leave him in that position. So I called him up and said, "Okay, Red. I'll take the job."

He said, "Good! You still going to play?"

"Oh yeah. That's the point."

"You made the right choice. Who better to motivate Bill Russell than Bill Russell?" He had told me the same thing many times in different situations, but never was it more apropos.

When Red and I had started to discuss my becoming coach, there were some things we didn't have to say. For example, when I was finally named publicly, I didn't know that I had just become the first African-American coach in the history of major league sports. Everybody was talking about breaking the color barrier, and all this "Jackie Robinson of basketball" stuff. Yet from the moment that Red and I agreed I'd be coach, nothing about race ever came up in our conversations.

It did not cross our minds. In fact, the first hint of race appeared in a newspaper headline the next morning: "Russell First of His Race." The first time it came up to Red and me was at the press conference in the Hotel Lenox to announce the change officially. After Red introduced me, I stepped up to take questions, and a reporter said, "You're the first Negro coach of a major league sport. Can you do the job impartially, without any racial prejudice in reverse?"

I said, "Yes!"

The reporter asked, "How?"

I said, "Because the most important factor is respect. And basketball respects a man for his ability, period."

When the media interviewed Red later, everyone pressed him about the social ramifications of having the first "Negro" coach. He was rightfully annoyed. "Look," he said impatiently, "it's no big deal. I just did what I thought was best for our team. If it was about anything else, I wouldn't have offered Russell the job. Or he wouldn't have taken it."

Considering the raw atmosphere, Red's evenhandedness was a great comfort to me because a lot of Bostonians were opposed to my becoming coach. Particularly the writers. When I took the job, one reporter wrote seven articles focusing on why I shouldn't be coaching the Celtics. Another one asked me on a TV show, "Now that you're part of management, has Red taken you into the woodshed and told you that you have to start signing autographs?"

I said, "There's something I want you to understand. Besides the fact that I'm all grown up, Red knew who I was before he offered the job to me. I didn't get the job and say, 'Now that I got it, I won't sign autographs anymore.' If I tell you I'm not going to sign autographs, I won't be signing autographs. And there's nobody you can appeal to who can tell me I have to do that. That person does not exist. In fact, not even God Almighty can make me do that. I don't understand this obsession with me signing autographs. Why don't you ask me about coaching the Celtics?"

Another source of public discomfort was the fact that I was supportive of, and outspoken about, civil rights causes nationally. I took part in a memoriam for Medgar Evers at Boston Common that attracted notice. I received the Crispus Attucks Award in Boston for my activism. It rubbed a lot of Bostonians the wrong way. At the time, Boston was a totally segregated city—and I vigorously opposed segregation.

For example, in June 1963, there was a huge uproar over segregation in Boston neighborhood schools. There were thirteen inner-city schools that were at least 90 percent black, and these schools were not nearly "separate but equal," let alone integrated. So the Boston chapter of the NAACP demanded a public acknowledgment of this overt segregation.

A woman named Louise Day Hicks, who was on the Boston School Committee, refused to admit publicly that any segregation existed in Boston schools. One day, she came to give a graduation talk at Roxbury Junior High School, where it clearly did exist. But when a preacher in our little ad hoc committee of activists made a fuss, and she couldn't finish her talk, she had him arrested. He told the press later, "Louise Day Hicks speaking at Roxbury Junior High School is like Hitler speaking in a synagogue."

Meanwhile, those kids didn't have their graduation. So we hosted a special ceremony for them at a local church, and I gave the commencement speech. After that, some folks tried to reach me with the usual racial epithets. They knew nothing about Bill Russell—they had no idea that I could not be abused. Afterward, however, I was not very popular with a lot of Bostonians, especially as the choice for Celtics coach.

Much more significant to me was the fact that none of this was ever a topic of discussion between Red and me—it just wasn't part of our relationship. His never asking me about my attitudes on civil rights, or being black in America, was the same as my never asking him about Rosh Hashanah, or being white. That wasn't because we didn't care or were deliberately looking the other way. It was just that I came from my tribe, and this was the way *I* arrived, and he came from his tribe, and this was the way *he* arrived. I don't think Red ever had a complete understanding of my views on race, but he respected them because he respected me. I knew he recognized that the things he did in his psyche to be a Jew, I was doing in my psyche to be a black man.

People had tried to bait him the same way they did with me, and our reactions were the same. He would not let anyone else tell him what to do—his role model for this would come from his tribe. That was how I viewed it. And nothing could move me off that view. That was why, when that reporter had asked me if I could coach without prejudice, instead of taking offense, I said, "Yes!" emphatically, and left it at that. Yes was all I *had* to say. The rest I had to *do*.

Still, a lot of folks assumed that I was another angry black man suffering under the boot of white oppression. They were mistaken. I *never* allowed myself to be oppressed. To me, it was like something Howard Cosell said when he was under fire for his work on Monday Night Football. He once said, "That's like lobbing spitballs at a battleship! They can't hurt me!" In fact, I can truthfully say that I was having the time of my life, even in dealing with the overt racism that vented my way. The joy was in finding the right ways to deal with it.

For example, raccoons are the smartest of all the wild animals. How do I know? Well, in 1957, I bought a house on Main Street in Reading, a Boston suburb. Of course, people knew that I was a Celtic, which meant they also knew when I was on the road. When I returned home after our first road trip that season, my wife told me there had been some vandalism while I was gone. The trash cans had been overturned and garbage was everywhere, so I had to clean it up. Next road trip, the same thing happened.

Being a good citizen, I visited the police station and reported the incident. I asked if I could call in to let them know when I'd be on my next road trip, so they could patrol our vicinity more frequently. The captain said, "Oh, that won't be necessary. It's probably just the raccoons." I said, "Okay. And by the way, is there a place I can get a gun permit? Because I want one." That was when I found out how smart raccoons were. I didn't even get a gun, and yet those raccoons heard that I'd gotten a permit and they never messed with my trash cans again!" I enjoyed the hell out of that.

Red communicated the same attitude to us on the Celtics: *Let's go find ways to win and enjoy the hell out of it!* Which we did. That was the passion exuding from all of us after we won another championship. Red and I expressed this in a motto we came up with on our basketball journey: "Play like children without being childish." It applied to our lives as well. Whenever someone got out of line with me off court, my thinking wasn't "How do I get my revenge?" It was, basically, "How can I take control?" Or, as Red used to say, "It's water off a duck's ass. No dampening of *these* feathers."

The point is, all these social matters were irrelevant to Red and me and how we conducted our relationship. We were as unconcerned about the public's view of my social activism as we were about making history when we had started five black guys for the first time. Add to that the fact that the writers all knew that I didn't give a damn about what they said about me. We concluded that the same people who didn't like Red for Coach of the Year—mostly because they mistakenly judged him as too arrogant—also didn't like me. We didn't care. "Not a big deal," we both would have said.

I sensed the stress lying in wait for me as "the first Negro coach" instead of "the Celtics' next coach." But I remembered that when I had left Red's office while we were still calculating who the new coach would be, an image popped into my mind of the last scene in *Casablanca*, where Bogart's walking in the fog with the inspector, and he says, "This could be the start of a beautiful friendship." And something encouraging occurred to me: "I don't know what'll happen with the new coach. But now my relationship with Red is just about friendship."

So I started looking forward to that.

There were many subtle ways that Red and I expressed our respect and affection for each other as men. By the time I was coach and he was general manager, it happened frequently. For instance, early in my second year as player-coach, I noticed that something was bothering our young star John Havlicek. He never voiced displeasure or called attention to himself, so I knew it was important. I asked him, "What's wrong with you, John?"

He said, "I'm pissed off. But it's nothing three thousand dollars won't cure."

"What is it?"

"Well, I just signed a contract with Red, and he browbeat me into signing for less than I wanted. I wanted three thousand more, and he wouldn't give it to me. I'm really upset."

I went right to Red and said, "What're you doing with Havlicek's contract?"

"Russ, you can't take the player's salary position against management now. You're management."

"That's bullshit, Red. You want me to send one of my best players out there pissed off, right before the season starts? I want to ask you a quick question then. Would I do that to you?"

He thought it over. "What do you want me to do?" That was how we always solved a potential disagreement: *How can I help?*

I said, "Here's what I want you to do. I want you to call John in and tell him, 'Listen, I browbeat you into signing a contract and you didn't want to do it. I was wrong. I shouldn't have done that to you.' Give him three thousand more. But I don't want you to tell him I told you to do that."

"Why not?"

"Because I'll be leaving one day, and it'll be very difficult on the next coach. I want John's loyalties to be to the Celtics, not to me."

Red relented and did what I asked. This small event, like so many others, was a reminder of how simple friendship can be. It demonstrated the kind of trust and respect we had for each other's integrity and judgment, which was a big reason why we never had a major disagreement. And that was a major reason why our relationship lasted so long.

For the next season, when I would be turning thirty-five, Red offered me an unprecedented eight-year, "no cut" contract. Every year of the contract called for the same salary. It was twice what I had ever made before. Red had also designed it so that I would collect whether I played or not. Each season, all I had to do was say, "I want to play this year" and show up for training camp, and that triggered the contract. I could even bring my own doctor to camp, and he could say, "Mr. Russell wants to play. But I don't think he should," and they would still pay me for the year. I would never have done that, but that was how Red wrote it up for me.

He didn't have to do it. But we both knew it was his way of giving me a "payback." When I asked him, "Red, what's this for?" he didn't admit that, of course. He just said, "There was a time when this franchise was losing money, and you played for less than you should have been paid. We're just catching you up."

He was right about the pay issue. In the early years, around contract time, Walter Brown actually told me, "We're not paying you enough. I know it and you know it. But I want to show you the books because we're losing money. That's why we can't afford to pay you more than we're paying you." And he showed me the real books, although I had no idea what I was reading. I thought that gesture was magnanimous.

I said, "Walter, come on. You're one of the few people in the world that I trust completely. All you had to say was, 'We can't pay you more this year because we're losing money.' I would've said, 'Okay. Pay me when you can.'"

So I signed a contract that season for a modest raise—nothing that would hurt him. I did it for a few lean years because that man would have given me *his* salary if I had asked for it—wrapped in the shirt off his back. Red knew about that meeting, and I always thought it figured into the eight-year deal he gave me later. This was the good-guy side of Red, trying to do a little act of kindness. But not because the club was making money now. It was Red's way of doing something nice for his friend. As if to say, "Thanks, big guy, for all the great years."

He had tried something like this before. In the early 1960s, after I'd already signed my contract, Red called me into his office to offer me a $25,000 bonus for the coming year. I said, "For what? I haven't done anything yet."

"You're probably going to lead the league in rebounds again."

"How do you know that, Red?"

He tilted his head, raised an eyebrow. "I read it in my matzo fortune cookie. I *know* it."

"I won't take it."

"Why not?"

"Well, for one thing, our plays require all five guys. Sometimes a play requires me to be in the Four position. I can't get rebounds from there. That means if somebody else in the league is having a great year rebounding, and we go into the last month and we're equal, I'm so competitive, I might stay around the basket to get the rebound."

I was busting his chops. I knew he was just trying to do me a favor and, essentially, give me an extra twenty-five grand. He would never say, "You're my friend. I'm trying to do you a favor." But I knew it. I had recognized early in our relationship what almost nobody else saw: beneath all his harsh bluster, Red was a genuinely good guy, like his buddy Walter. The professional side of him was brash, loud, combative. But the private side, at least to me, was warm, sensitive, big-hearted. The public never saw that side of him. Just like our private relationship, it was not for public consumption.

You might have expected that once Red became general manager and I became coach, the whole dynamic between us would change. But it didn't change that much. In fact, our friendship only deepened. Red was really good with details, so if something in my coaching strategy wasn't working, I'd ask him to come to practice and consult about the details. I considered him the greatest basketball mind ever, and I knew that he loved keeping his hand in, especially if he could help me. I'd say, "You watch the game and tell me what you see." That harkened back to all those games when he deliberately had me sit beside him on the bench while I was resting, to watch the game with him through his eyes. His input now was equally helpful, which I appreciated. And he appreciated the continued involvement—because he never would have asked to be asked.

He always insisted that whenever we consulted, we do it in private. He didn't have to explain why. I understood that it was because he didn't want the players to ever think that I had lost authority. Or that—my words, not Red's—"the legendary Red Auerbach" was looking over my shoulder. He would not have wanted me to heft that weight. Only a devoted friend would have thought about that.

During those last three years of my career, 1966 to 1969, whenever I went to Red's office to talk to him about coaching, it was mainly about personnel matters, such as things I'd have to guard against. He said, "There's never enough minutes and never enough shots. But you can't let your personal prejudices determine the politics of who gets to play over anyone else." He knew, for instance, that Sam Jones had been my roommate before I was named team captain in '64, so he was hinting: "Don't be inclined to get Sam more shots, just because you like him."

We talked about that at length—it was like being in school. But it wasn't like a teacher talking to a student, or a general manager talking to a coach. It was just two good friends talking about their profession. That was a slightly different dynamic than when I was just a player. Moreover, as G.M., Red attended home games but didn't go on the road with us, so he wasn't there to observe everything and give me advice accordingly. I was the coach, and the only one who saw every game up close.

He enjoyed consulting like this for a few reasons. First, his agenda was still to win games, and he was just as passionate as ever to contribute personally. Second, it gratified him to help his friend become a better coach: "Okay, Russ," he'd say, "here are some of the things I looked out for when I was coaching. You can't do it the way I did it. We're two different people. But here are some things you can put in your guidebook." He would never say, "This is the way you should do it." And I sensed him being very careful about what he told me. He didn't want to overemphasize something that might have mattered more to him than to me. He negotiated that terrain as a friend who knew exactly what *I* thought was best for the team before he offered his views.

In 1968, my second season as player-coach, we won another championship. It was especially exhilarating because I had fulfilled my last basketball agenda: winning a title as a coach. One was enough, so I decided to retire. But when an important personal matter intervened, I changed my mind, for just one year. I kept this to myself but I knew, going in, that 1969 would be my last year in basketball. It was some year: we won the title yet again, number eleven for me in thirteen years. And yet, two days later, a reporter walked into Red's office and asked him, "As G.M., are you satisfied with the coaching you had this year?"

Red said, "What the hell are you talking about? We just won the goddamn *championship*!"

"But if you had a better coach, don't you think you would've won more regular season games?"

Red called me afterward and told me about this exchange. "Jesus Christ," he added. "Do you believe this dumb shit?"

I said, "Well, you know, it makes no difference, Red. I'm not doing this anymore anyway."

"What do you mean?"

"I'm quitting the game."

There was a big pause, but I knew what was coming. "Russ," he finally asked. "Did you tell anyone else?"

"Why?"

"Because you know I have to talk you out of it. If you tell anyone, it'll be impossible for you to back out of it."

"Red, you're the only one who knows. But I'm through. I'm retiring."

I half-expected him to try to talk me out of it, but he never did. From that moment on, he never once said, "I want you to come back out." He told me later that he didn't attempt to dissuade me because he felt I might lose respect for him.

When I got ready to leave Boston for good, there was no last meeting with Red to say good-bye. It never occurred to us it *was* good-bye. We were very close by then, and we knew that feeling of closeness would continue. To some people whose friendships are more traditionally "social," this might seem unusual, but it wasn't strange to Red and me. We were not social beings; we liked each other's company, but we were very comfortable in our own company, too.

There's a distinction between being alone and being lonely. Red always had a lot of very devoted friends—more than he ever wanted. And, in my small, select circle, I had all the friends I wanted. A New York reporter once wrote— accurately, for a change—that every time he saw me, I was by myself. Well, every time I saw Red Auerbach, he was by himself. That was fine with us. We knew we didn't need to spend time together to keep our relationship going.

Chapter 9

|||||||||||||||||||||||||||||||||||||||||||||||

This Is My Friend

When I finally left Boston and my career, I felt no sense of loss, no regret, no particular sadness. I left behind the basketball relationship, but not my friendship. Red and I both knew that would never change. With Red in D.C. and me settling eventually in Seattle, we didn't spend extended time together ever again. Yet, despite long separations and just a few phone conversations a year, there was never a vacuum. And nothing really changed in our connection.

I used to see Red in Boston when I came to visit my daughter at Harvard Law School, or to attend special events, or on business commitments. For ten years, I also played in the Member-Guest golf tournament at Woodmont, the country club that Red belonged to in D.C. Every time I showed up, Red was in the card room playing gin. I used to stick my head inside and ask, "How you doing?" He'd grumble, "You don't want me to come out there, do you?" Still the same Brooklyn needle: *Red Auerbach deems Bill Russell worthy of interrupting his card game to come say hello, but it better be good.*

"No! Jesus Christ, no, Red! Play your cards!"

There was nothing forced between us, no agendas to push, no sense of a special occasion. It was just business as usual. Same friendly sarcasm, same "no big deal" attitude, same Red, same Bill. The key was always the purity of the friendship. We didn't have to put out any extra effort. We didn't have to play "the friendship card" or embellish the relationship in any way. Our interest in each other no longer came from basketball. We were now, simply, two men. When you have that sort of old-shoe comfort with a friend, it's like having a pacemaker. Once it's synchronized, you don't have to think about it. It's just there, working for you. That's how our friendship always felt—even more so in retirement. It was just there, working for us, so we didn't have to think about it anymore.

When I poked my head in the Woodmont card room, it might have been a year, or more, since we'd last seen each other. We wouldn't have known or cared. The reason he was there was to play cards, which he loved; and the reason I was there was to play golf, which I loved. Although we both knew we were looking forward to seeing each other again, just saying hello and knowing we were there was sufficient.

Outside the club, we rarely socialized—that part didn't change either. In our "dotage," aside from the occasional special event honoring one of us or someone we respected, about the only thing we did in public together was attend some Celtics games. Afterward, we always had a quiet, private dinner at the China Doll, his favorite Chinese restaurant. But that was it. We both had our own ongoing lives—we knew the situation.

The rest of the time, we stayed in touch pretty regularly by phone, a comfort that was important to both of us. Most of those conversations were brief, like our basketball conversations during the Celtics years. We talked about ordinary things that everyone else discussed: what was going on in our lives; our health; our observations on current events; and, of course, sports. As always, I would tell him only what I wanted him to know about my current life, and he did the same. It was more about listening to each other now; that was how we learned where we were, at present, in each other's life. I always said, "I'm only interested in what's real. I'm not interested in bullshit." Red felt the same way. So, that's what we talked about: what was real, *to us*.

We always talked a little about what was going on with our families, especially our kids. When my daughter, Karen, attended Georgetown University in D.C., without being intrusive Red used to check around to find out how she was doing, to see if she had any problems or needed anything. He wanted to be there for her, you might say like a grandfather. But he never told her that. I knew this because he'd call me up and say, "I just checked up on your daughter again and she's doing okay." He didn't know it, but I was deeply flattered. When his daughter Randy moved to L.A. and became my daughter Karen's friend, I checked up on her, through Karen, and I let Red know the same thing: "I checked on your daughter again and she's doing okay." No big deal. Just mutual acts of kindness between friends.

As Red and I grew older, we got together less and less. Yet the special feel we had for each other, and the way we learned to give and accept little kindnesses or help from each other, continued to ripen. In the 1970s, when Red arranged to retire my number up to the Boston Garden rafters, I told him I would not take part in a public ceremony for that. My reasoning was, first, that I had played for my team, not for personal trophies or honors. And second, the last thing I wanted for myself was a public fuss over Bill Russell, the basketball player. That was why I left town in 1969 very quietly, without any of that public "Good-bye Bill" fanfare.

It reminded me of the time I read an obituary for a woman who'd been dead a year, and I thought, "How cool is that!" I figured she didn't want any fuss. That was how I felt when I retired. When I was still playing, all these players who had announced "This is my last year" were being honored at ceremonies in different cities. I remembered Red and me standing together at Bob Cousy's retirement ceremony in 1963, at Boston Garden, as people yelled from the stands, "Don't leave!" and "We love you, Cooz!" I thought of the taunts some of these same fans had hurled my way in my career, for speaking my mind, such as when I said, "I don't play for Boston. I play for the Celtics" and "The fans don't know anything"—which I did not consider disrespectful. I thought, "I'm supposed to miss those folks? Hell no! I won't miss them!" So I turned to Red and said, "Boy, when I retire, I will never do *this*!"

I reminded Red of that now. "That's bullshit," he said predictably. "I have to do this, Russ. And there's just some things you have to do, too. It's for the organization. You have to be there when your number goes up!"

"Red, if you really want to honor me, you will not retire my number, because that is not what I want."

He said, "Well, you got to."

Stubborn versus Even More Stubborn. We went back and forth on it until one day, when I arrived at the Garden to broadcast a TV game, Red confidently informed me that my number would be retired that night at the game, and I was supposed to stand up when they announced it. He had ambushed me because he knew I had to be there anyway, so I'd be a captive audience. I said no again. "What about the broadcast?" he poked around. "You skipping that, too?"

Finally, I said, "Okay, Red. Tell you what. I played for my team and for my teammates, not for my own glory. You know that. But if you insist on retiring my number, just sew it on the flag with the other numbers and run it up there. But no public ceremony. I'll do it if you put the jersey up with only my teammates present."

Grudgingly, he said okay. So, that night, before the Garden opened and the fans arrived, Red and the Celtics players still on the team from the time I'd played stood in the middle of the floor and raised my number before 13,909 empty seats. It was a somewhat imperfect solution. But the important thing was that Red and I still understood each other well enough to find one.

Over twenty years later, in May 1999 at the Fleet Center, the Celtics' new arena, there was a benefit tribute for the National Mentoring Partnership and the Massachusetts Mentoring Partnership—programs dear to my heart. Afterward, there was a special ceremony for the re-retirement of my number. I was embarrassed, but I had agreed to the ceremony because I felt it would provide worthy exposure for those wonderful organizations, and for mentoring in general, something I believe in deeply.

When the moment arrived, I noticed that the large, white-and-green banner wasn't just flying my number. It had all the previously retired numbers of my teammates. That was a huge relief—and, I felt, still the perfect way to honor me:

as part of my team. Red and I stood together on the famous parquet floor that they'd transplanted from the old Boston Garden, and we both took hold of the rope. Just as we started to tug, we winked at each other—our whole friendship contained in those winks—and started hoisting our team's banner together, hand-over-hand, to the rafters. I will always remember that feeling as something very special.

There was a sweetness to it that recalled something from my youth. I never had a best friend growing up—that was just the way it happened. But my father had one friend for sixty-six years. They had walked to first grade together. When they were both in their seventies, they were still friends—an unbroken connection. In fact, after my father got settled in California, this guy moved to California, too, so they were separated for less than six months their whole lives. My father had two or three other close friends, and it was wonderful for me, as a kid, to see them all together. It was almost as if my father had a glow—you could see it was special for him to be with these people. And no matter what the situation, these guys were always good with each other. My father was a big, strong man—six foot three, 212 pounds—and yet, when he was with these friends, there was a tenderness, or a kindness, they showed each other. I don't know that I understood it as a child, but now I do: this tenderness came from inner strength. It was something you didn't see much between men back then. And men didn't talk about it. I sensed it then. And I sensed it again, hoisting that banner with Red.

On the other hand, I was thinking, "I told you I would never do this in front of a crowd. So you won *this* one, damn it." And he was probably thinking the same thing: "I got you *this* time."

Sometime after 2000, Red told me, for the first time, that it really hurt him when the Celtics caved in to new coach Rick Pitino's demand to be named president. That had been Red's title since 1984, along with general manager. "Pisses me off," he said. "It hurt my feelings. We went to war together so many times. And then to do this to me after all we've been through!"

I said, "You don't need that title, Red. You got a nice retirement. You'll never need money or anything else. Why the hell should you care about that?"

I'll never forget what he said: "As long as I'm president of the Celtics, people return my phone calls." He was serious— even though the job was mainly ceremonial, it still mattered to him. That was partly his pride and partly his deep loyalty, especially after all he'd done for the club. I understood his feelings completely. The title was like a Congressional Medal of Honor to him: "I've earned this."

I said, "The only conclusion I can draw is that those people are full of crap." Meaning the ownership and Pitino. "If Walter Brown had been alive, no way in the world that would've happened."

"I agree with that!" he said.

After that exchange, we never talked of it again. I always said that, as my life progressed, I didn't want to dwell on a list of grievances. Red felt the same way. This was the only thing in his Celtics career that still wounded him, and he knew that I would understand. After nearly fifty years, it still touched me that I could do something to help him, even if it was just to listen and understand. We could share moments like these now—they were more intimate than anything we shared on the Celtics. That was part of the evolution of the relationship from professional to personal. Sometimes, it comes to you like fate—suddenly you just *have* it. Other times, you "find" it. A few years ago, I was with a group that had a private audience with the Dalai Lama. He spoke to us and then asked if anyone had a question. I said, "I would like to know something. When did you reach this state of serenity, as I call it, when you were able to reconcile spirituality with reality?" He said, "I reached my serenity over a very long time, as I evolved as a person through the teachings I received as a child. There was no one day where I said, 'I *have* it!'"

The same was true of my friendship with Red. It had evolved over a long period of time, through our experiences together and everything they taught us, until we reached a certain serenity within it.

The most comfortable moments that Red and I spent together after we both retired were at Celtics games. Our dynamics carried over from the past. As with most of our encounters, he was by himself and I was by myself. We rarely spent time together with anyone else. In a strange way, that pattern continued, physically and psychologically, even in an arena jammed with 20,000 people.

Red always sat in Row 7 and I sat in Row 8, right behind him. If we had sat beside each other, we would have had to look sideways to talk, and we couldn't watch the game. We didn't discuss this or plan it—it just happened one time, and we both realized it was the way we wanted it. This way, I could lean over and talk into his right ear, and we could both keep our eyes on what was going on down on the court.

We really enjoyed watching the game and exchanging comments. The last ten years that we attended games, he always complained bitterly about the loud atmosphere. What we both loved about basketball was still the same thing: the competition. But he thought that was diluted now by the incessant thumping music, the flashing video displays, and what he called "the whole bunch of other shit going on" that had nothing to do with the game. He especially hated seeing the female dancers—he said they looked like a circus act. At some point, he'd always yell, to no one in particular, "Hey! There's a goddamn *game* going on!" He thought all these distractions pandered to the "local yokels." Apparently, they were still hanging out at Celtics games.

I loved his energy and passion, and how deeply he still felt about the game that we had helped to revolutionize together. And I loved the easy way we related to each other now, as just a couple of guys. But there was a moment that I will always remember above anything else at all those games. It happened in his later years, when he was growing very frail. As we watched the game, and schmoozed and needled casually, enjoying our now very precious time together, I had the sensation that my presence, behind him and above, made him feel safe. And I admit that, for one instant, I felt that I was still protecting him, the way I always had as a player.

In a DVD that the league produced about us, there's a shot of Red and me at one of the last games we ever saw together. I'm leaning forward, and he's watching, and we're having a good time. If you look closely, there is something in our expressions that says, "These two men are brothers. And the one behind is watching over the one below."

But that's probably just my imagination.

||||||||||||||||||||||||||||||||||||||||||

Good-bye to My Friend

My friend Red Auerbach's heart gave way on Saturday, October 28, 2006, in Bethesda, Maryland, just after his eighty-ninth birthday. It was almost fifty years from the day I joined the Celtics. His daughter Nancy phoned me in Seattle. "Bill," she said, "we just lost Daddy. He died a couple of hours ago. I didn't want you to hear it someplace else. I wanted you to hear it from us." She said that only three days earlier the Navy had presented Red with an award for service to his country and his lifetime achievements. He made a nice speech, and enjoyed himself, and was really upbeat. And now he was dead.

I just listened. I knew that, in the Jewish culture, burial happens fast—it's a short cycle. So the first thing I said was, "When are you having the funeral?" I was thinking about my travel arrangements. Nancy said she'd call me later with the details. I said,

"Okay. My thoughts are with you two girls and your families." And we hung up.

Nancy calling me within two hours of her father's death confirmed how sensitive his family was to our friendship. I didn't know that before then; it was never discussed. She'd said they lost "Daddy," not "Red"—as though I were one of the family. The fact that they knew how much we cared about each other was very gratifying to me. His wife, Dorothy, had been gone six years. So now there were only Nancy and Randy.

I had known them since they were young. I met Nancy, the older child, just before I left for the Melbourne Olympics. They adored their father—they felt the same about him as my daughter, Karen, feels about me. They knew I had enormous affection for them, not just for their father. It was funny, because Randy, who was a friend of Karen's, used to tell her, "My daddy just loves your daddy!" And Randy told me things like, "Daddy's crazy about you!" That was something Red had never told me. And I wouldn't have wanted him to.

Right after Nancy's call, I felt a tremendous sense of loss roll over me. Red and I had last talked on the phone the week he died, and now we wouldn't ever talk again. It wasn't totally unexpected. I knew that after Red's colon cancer surgery in 2005, and some tough respiratory problems, his health had run downhill. He struggled, though he never said anything to alarm anyone or to elicit sympathy. That wasn't Red—he would never want anyone's pity. Yet I was still surprised. And disappointed. We think that our friends are immortal and will always be there. When they get sick, we know they can die, but we never expect them to be gone, just like that. Death is one of those possibilities we expect only vaguely, and we hope it doesn't happen.

Immediately, I started thinking about the funeral. I knew it would be a pretty big event, with people wall-to-wall. I thought, "Red would've called it a circus. He would've hated it." To me, funerals are for the living, not the deceased. They're mainly for family and friends. I had two reasons for going: I wanted to be there for Red's girls, to be of comfort to them, and I wanted to say good-bye to my friend. That was my agenda—it was deeply personal. Circus atmosphere or not, I'd just have to find ways to negotiate the crowd.

Many different ingredients go into making us what we are. One of the most important is true friendship. True friends occupy special places in our psyches. They don't have to interact with us every day, or every month, or every year. They're always a part of us. It's not that we take them for granted. We just incorporate them into our lives so naturally, we don't realize they're there. It's when they're gone that we realize what we've lost.

When I left for Washington the next day, I started to really feel my loss. I thought, "He's really gone. I can't growl at him anymore. And he can't growl at me." It was a long flight across the country, so I had a lot of time to reflect. I thought about our last talk: "I hope we had the right last conversation. Did we convey how much we meant to each other? Did we leave anything unsaid? Maybe we left some things unsaid." Then I thought, "No. He knew how I felt about him, and I knew how he felt about me."

My attitude, in difficult situations, is to keep moving forward positively, so my own mortality never crossed my mind. To me, death is not a tragedy. We're born, we live, we die—it happens to all of us. That's the way it is. Years earlier, at my father's funeral, I had thought of a Glenn Yarbrough song called "Grandma's Letter." It's about a grandmother's note that her son reads to the family after her death. When Yarbrough sang, "Guess from here on, we're the old folks now," I felt that way myself. But even then, at my father's service, I stopped thinking about myself and I remembered the wonderful things my father did for me as I grew up that helped to make me a man, and how much I had loved and respected him my whole life.

Nowadays, when someone near and dear to me dies, the first thing I wonder is, "Was he content with his life?" To me, that's the sign of a life well lived. I knew that Red felt that way about his life—it was very full. He was at peace right to the end. Another thing I try to think about is anything funny we did or said. In the plane, on the way to Red's funeral, what came to my mind was the night, early in my career, when Red and I were playing gin rummy and talking basketball. All of a sudden, he said, "Tell you what I'll do, Russ. At this stage of the game, it's time I gave you a secret name. We'll both use it, just for you."

"Okay. What's that, Red?"

"It's a good one. It's perfect for you. You'll thank me for it someday."

"Red! What *is* it?"

He rolled his cigar around his mouth, arched his eyebrows, and said, "*Goyishe kop!*" I started to laugh. I knew he was needling me about how I played gin because he always kicked my ass. So I gave him a scowl. He looked at me seriously and said, "Wait a minute. You know what *Goyishe kop* means?"

I said, "Yeah, I do. Means '*Dumb head*'! And thanks a lot!"

He chuckled. "Okay, I was just giving you shit." And we both cracked up. Recalling this amused me all over again. I could see Red's mischievous smile. And that brought to mind an even funnier incident, when Red was caught off-guard, which almost never happened. It was the time he gave Tommy Heinsohn—one of the Celtics' notorious characters—an exploding cigar. Tommy was having a lousy day. He was in the process of a messy divorce, and he and his wife were yelling at each other, so he was late driving to practice. He knew that Red would be all over his ass, so he started speeding, and a cop pulled him over and gave him a ticket. Tommy told himself, "Screw it. I'll be late anyway. I'll just sit here and smoke a cigar and relax. Then I'll go to practice and I'll be able to handle anything Red comes up with." He lit a cigar—but it was one that Red had given him, and it exploded. Finally, he arrived at practice, soot all over his face, and Red laughed like hell.

Tommy spent most of the next two years presenting Red with real cigars. But Red always had Tommy puff them first, in front of the team, as a test. He was too smart—he knew that Heinsohn was plotting revenge. Finally, after dozens of good cigars, Red dropped his guard and quit the testing. So, one day at practice, Red was addressing the team about something important when Tommy slipped him the loaded cigar. Red lit it up . . . and *Bang!* It caught him so much by surprise, he was beside himself. He never saw it coming. Especially from Heinsohn, who was not known for his infinite patience.

That was how we got along on all of Red's teams. It wasn't, "I'm the coach and I run the show, so I'm untouchable." It was, "We are all one unit, and in this unit, I do the coaching and you do the playing, but there's no hierarchy." That was why a Tommy Heinsohn could get away with slipping Red Auerbach a trick cigar. The smoke always cleared by the very next day.

Picturing all this again, I almost laughed out loud. I could see Red's face when the cigar went off, and I could hear his laugh. Then I thought, "The thing is, Red never would've accepted an exploding cigar from *me*, even if I waited ten years. He always saw me coming. But now I'll never get the chance to try."

When Marilyn, Karen, and I arrived at the funeral home, a huge crowd had already gathered—players, coaches, friends, and Celtics personnel. It looked like the whole city had shown up. There were TV reporters everywhere, extending their microphones rudely to interview everyone about Red. When they asked me, I said, "No. I won't do that," and waved them off. My friend had died; I wasn't there to talk publicly about our relationship. I didn't want to talk to anyone, period. To avoid any more attention, we walked around the side of the building and entered through the back door.

We came into a large room where the casket was on display and made our way to Nancy and Randy, who were greeting well-wishers. We all hugged, and then I nodded at a couple of my ex-teammates. When the service got under way, Marilyn and I took seats in a corner, off by ourselves, as far away as possible from the goings-on. There were plenty of eulogies—a lot of people had a lot to say. I didn't pay much attention because I wasn't interested in eulogies. Red was my friend; I knew all I needed to know.

But I picked up fragments, and it occurred to me that none of the speakers knew the same Red Auerbach I knew.

And that he would have hated these tributes. He was uncomfortable and impatient with big shows of emotion, especially over him. I could picture him sitting up in his casket and yelling, like he did at the referees during our games, "Jesus Christ! What the hell's the big deal? Let's get the goddamn show on the road!"

I waited until everyone finished their speeches and people started filtering out. At the right moment, I walked toward the casket. It was closed, and I was glad about that. I prefer to remember my friends the way they were when we were together. When I approached the casket, I thought of the last time I saw Red, when he said, "Don't fall," and how earnest he had looked when he said it, and how that friendly warning had touched me. It was so vivid, it is still engraved in my heart.

I stood alone at his casket. No one else was close by. I made sure of that, because this moment was intensely private. "Just like when he coached me," I thought. "Always one-on-one with nobody else around." I reached down and set my hand on the lid for just a second, and I thought, "Good-bye." It was the same thing Red would have done for me.

All of us have this dark place inside us where we don't allow anyone else to go. Yet all our lives, we seek to let someone get a glimpse of that place and maybe even reach inside and touch us. Just a touch—any more than that would be too much to bear. The closer the friendship, the more often they get a glimpse, but it always remains our private sanctuary.

At Red's burial, as they started lowering the casket, I was thinking about how lucky it was that he understood and respected this boundary. All of a sudden, I remembered telling Red one time that it would be perfectly all right with me if I didn't have a funeral and I was buried in an unmarked grave. He'd looked at me sideways and said, "Come on! You're full of it!" In all the years I knew him, he was right about almost everything. But he was wrong about that. Because my view is that we are all equal creatures of God. And when we die, *He* doesn't need a marker to find us.

I decided right there that I would not mourn Red's death. Instead, I would celebrate his life. And my special place in it. And his special place in mine. And this unique friendship that lasted half a century.

The next day, back in Seattle, the sky was deep and clear, so I played golf as usual. When I stepped up to the first tee, it occurred to me that if I had died before Red, and he came to Seattle for my funeral, when he returned home the next day, he'd have been sitting down to play gin at the club, smoking a cigar, and grousing, "Jesus Christ! Can we get the goddamn show on the road!"

And I wouldn't have had it any other way.

Acknowledgments

||||||||||||||||||||||||||||||||||||||||||||||

To Nancy and Randy: you both know how much I loved your father. This book is an acknowledgment of how much fun we had working together and being together as friends. And to my father, who prepared me for the encounter. While working on this book I lost my precious Marilyn.

—*Bill Russell*

For their help and support, I am grateful to my family, Debra, June, Michael, and Jean.

—*Alan Steinberg*

The authors wish to especially thank Bruce Nichols, Flip Brophy, and Peter McGuigan.